Research Report

Climate and Readiness

Understanding Climate Vulnerability of U.S. Joint Force Readiness

Katharina Ley Best, Scott R. Stephenson, Susan A. Resetar,
Paul W. Mayberry, Emmi Yonekura, Rahim Ali, Joshua Klimas,
Stephanie Stewart, Jessica Arana, Inez Khan, Vanessa Wolf

Prepared for the Office of the Under Secretary of Defense for Personnel & Readiness
Approved for public release; distribution is unlimited

NATIONAL DEFENSE RESEARCH INSTITUTE

For more information on this publication, visit **www.rand.org/t/RRA1551-1**.

About RAND

The RAND Corporation is a research organization that develops solutions to public policy challenges to help make communities throughout the world safer and more secure, healthier and more prosperous. RAND is nonprofit, nonpartisan, and committed to the public interest. To learn more about RAND, visit www.rand.org.

Research Integrity

Our mission to help improve policy and decisionmaking through research and analysis is enabled through our core values of quality and objectivity and our unwavering commitment to the highest level of integrity and ethical behavior. To help ensure our research and analysis are rigorous, objective, and nonpartisan, we subject our research publications to a robust and exacting quality-assurance process; avoid both the appearance and reality of financial and other conflicts of interest through staff training, project screening, and a policy of mandatory disclosure; and pursue transparency in our research engagements through our commitment to the open publication of our research findings and recommendations, disclosure of the source of funding of published research, and policies to ensure intellectual independence. For more information, visit www.rand.org/about/research-integrity.

RAND's publications do not necessarily reflect the opinions of its research clients and sponsors.

Published by the RAND Corporation, Santa Monica, Calif.
© 2023 RAND Corporation
RAND® is a registered trademark.

Library of Congress Cataloging-in-Publication Data is available for this publication.
ISBN: 978-1-9774-1045-0

Cover designer: Jessica Arana.

Limited Print and Electronic Distribution Rights

This publication and trademark(s) contained herein are protected by law. This representation of RAND intellectual property is provided for noncommercial use only. Unauthorized posting of this publication online is prohibited; linking directly to its webpage on rand.org is encouraged. Permission is required from RAND to reproduce, or reuse in another form, any of its research products for commercial purposes. For information on reprint and reuse permissions, please visit www.rand.org/pubs/permissions.

About This Report

This report documents research and analysis conducted as part of a project titled *Estimating Impact of Climate Change on Force Readiness*. The purpose of the project was to identify and build analytic capabilities to describe the relationship between selected climate hazards and joint force readiness in selected theaters of operation. The full dataset of narratives that describe the ways climate hazards can affect readiness is presented in a separate online appendix to this report.[1]

The research reported here was completed in November 2022 and underwent security review with the sponsor and the Defense Office of Prepublication and Security Review before public release.

RAND National Security Research Division

This research was sponsored by the Office of the Under Secretary of Defense for Personnel and Readiness and conducted within the Personnel, Readiness, and Health Program of the RAND National Security Research Division (NSRD), which operates the National Defense Research Institute (NDRI), a federally funded research and development center sponsored by the Office of the Secretary of Defense, the Joint Staff, the Unified Combatant Commands, the Navy, the Marine Corps, the defense agencies, and the defense intelligence enterprise.

For more information on the RAND Personnel, Readiness, and Health Program, see www.rand.org/nsrd/frp or contact the director (contact information is provided on the webpage).

Acknowledgments

We are extremely grateful for the support we received from Kimberly Jackson, Deputy Assistant Secretary of Defense for Force Readiness, and her team. We would especially like to thank George Landis, Lt Col Veronica Reyes, and Alejandro de la Puente for the many hours they spent helping us make connections and find needed materials. We also thank Lt Gen Richard Nugee and Sherri Goodman for helping us stress-test our framework. Finally, we would like to thank our reviewers, Michael McNerney and John Conger, and our leadership in the National Security Research Division's Personnel, Readiness, and Health Program, Molly McIntosh and Daniel Ginsberg. Any errors are our own.

---

[1] Katharina Ley Best, Scott R. Stephenson, Susan A. Resetar, Paul W. Mayberry, Emmi Yonekura, Rahim Ali, Joshua Klimas, Stephanie Stewart, Jessica Arana, Inez Khan, and Vanessa Wolf, *Climate and Readiness: Understanding Climate Vulnerability of U.S. Joint Force Readiness: Climate Hazard Pathways Appendix*, RAND Corporation, RR-A1551-2, 2023.

Summary

The Intergovernmental Panel on Climate Change's (IPCC's) Sixth Assessment concludes that global temperatures will undoubtedly increase at least through 2050.[2] Accordingly, the physical environment in which the U.S. Department of Defense (DoD) must operate is being affected by climate hazards and is thereby adversely affecting the performance of the joint force and the systems that support it. Generating, maintaining, and even increasing force readiness in light of changing climate threats is a key component of meeting the United States' highest-level strategic goals, from defending the homeland to deterring aggression and strategic attacks. The Office of the Secretary of Defense (OSD) identified a need to establish both qualitative and quantitative links between climate effects and readiness impacts. Acknowledging that climate effects are likely to become more severe as global temperatures rise, in this report, we document the results of an initial study aimed at developing such links between climate and readiness, laying the groundwork for the eventual integration of climate risk with quantitative readiness assessment and decisionmaking to help ensure that military forces can reliably and affordably sustain needed readiness in a changing climate. This study builds on previous RAND Corporation studies on installation resilience to natural hazards,[3] as well as on numerous efforts driven by Congress and DoD leadership to integrate climate change considerations within operational and infrastructure planning.

To our knowledge, there has not yet been a systematic effort to capture, organize, or model the complete landscape of potential climate change effects on joint force readiness. A key contribution of this study is a *climate readiness framework* for understanding the risk to readiness that may result from a combination of climate hazard exposure and the underlying vulnerability of a variety of readiness inputs to that exposure. The framework, summarized in Figure S.1, illustrates how *climate hazards*, which sometimes lead to causally important *second-order effects*, lead to *readiness impacts* intended to provide some information on the mechanism by which climate might disrupt relevant *readiness inputs*. This structure allows for standardization of language and application of a consistent organizational structure to linkages from climate to readiness, and it identifies 13 climate hazards that may lead to impacts on four readiness inputs: *people, training, equipment,* and *force projection*.

We populated the framework with 110 *climate hazard pathways*, obtained through a literature review and outreach with DoD experts and stakeholders, each telling the story of a particular way in which climate could potentially affect one or more readiness inputs.[4] The pathways include both documented cases of climate-driven hazards affecting one or more inputs to readiness and plausible mechanisms by which such

---

[2] IPCC, "Summary for Policymakers," in Hans-Otto Pörtner, Debra C. Roberts, Melinda M. B. Tignor, Elvira Poloczanka, Katja Mintenbeck, Andrés Alegría, Marlies Craig, Stefanie Langsdorf, Sina Löschke, Vincent Möller, et al., eds., *Climate Change 2022: Impacts, Adaptation and Vulnerability: Working Group II Contribution to the Sixth Assessment Report of the Intergovernmental Panel on Climate Change*, Cambridge University Press, 2022, p. 10.

[3] Beth E. Lachman, Susan A. Resetar, Nidhi Kalra, Agnes Gereben Schaefer, and Aimee E. Curtright, *Water Management, Partnerships, Rights, and Market Trends: An Overview for Army Installation Managers*, RAND Corporation, RR-933-A, 2016; Beth E. Lachman, Anny Wong, and Susan A. Resetar, *The Thin Green Line: An Assessment of DoD's Readiness and Environmental Protection Initiative to Buffer Installation Encroachment*, RAND Corporation, MG-612-OSD, 2007; Anu Narayanan, Michael J. Lostumbo, Kristin Van Abel, Michael T. Wilson, Anna Jean Wirth, and Rahim Ali, *Grounded: An Enterprise-Wide Look at Department of the Air Force Installation Exposure to Natural Hazards: Implications for Infrastructure Investment Decisionmaking and Continuity of Operations Planning*, RAND Corporation, RR-A523-1, 2021.

[4] Because of the large and complex nature of the tables containing the pathways, and to facilitate such functions as sorting and filtering, the full dataset of narratives that describe the ways climate hazards can affect readiness is presented in a separate online appendix to this report (Katharina Ley Best, Scott R. Stephenson, Susan A. Resetar, Paul W. Mayberry, Emmi Yonekura, Rahim Ali, Joshua Klimas, Stephanie Stewart, Jessica Arana, Inez Khan, and Vanessa Wolf, *Climate and Readiness:*

FIGURE S.1
The Climate Readiness Framework

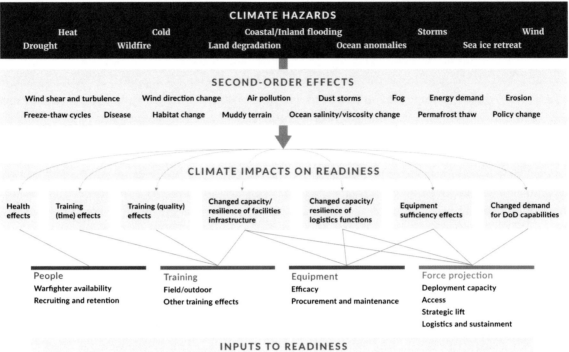

impacts could occur. This set of pathways is by no means exhaustive; it represents a set of possibilities rather than a dataset of all mechanisms. We also developed several diagrams and two interactive visualizations to help identify patterns and commonalities among the known climate hazard pathways. These patterns, in turn, can help users identify areas of risk that may require action to preemptively adapt to future impacts. In addition, we collected a set of *climate readiness integration points*—potential ways of incorporating climate risk into existing readiness processes, models, tools, assessments, and data structures—to be considered for further study or development by DoD and the services. Potential integration points include mission assurance processes; decision-support and planning tools; unit-readiness reporting and strategic-readiness assessments; existing exposure-analysis tools, such as the DoD Climate Assessment Tool; and under-development readiness prediction efforts being implemented in DoD's Readiness Decision Impact Model.

Findings

Climate and Readiness Are Understood and Studied by Separate Communities

Through discussions, we found that climate effects, especially over the long term, are not part of the readiness dialogue and that the readiness community is focused on more short-term, high-impact, traditional threats to readiness generation. Readiness operators are generally not familiar with climate terms and tend to be more focused on short-term weather effects, not yet fully considering the effects of future weather on future readiness. Furthermore, the relevant time dimensions for considering climate and readiness generation are different. Readiness leaders have immediate demands that often preclude them from dedicating

Understanding Climate Vulnerability of U.S. Joint Force Readiness: Climate Hazard Pathways Appendix, RAND Corporation, RR-A1551-2, 2023).

attention to potential challenges that do not have immediate impacts to operational and tactical outcomes, and longer-scale strategic planning may be the responsibility of other leaders. Climate will, however, have significant impact on the short-term readiness of the future; as climate shifts, the kinds of events that require attention from commanders could change in scale, frequency, and nature.

Commensurate with the difference in timescales, readiness-related decisionmaking occurs primarily at the operational level, but climate-related decisionmaking will need to occur primarily, although not only, at the strategic level. At the strategic level, responsibilities for decisionmaking and planning at the intersection of climate and readiness are not always clear. Climate is only starting to be included in broader, long-range planning by DoD and the services. Individuals with climate expertise are not typically present in decision forums, senior decisionmakers are not always being educated to present climate issues from a strategic and all-encompassing perspective, and there is no dedicated organizational structure within which to systematically analyze and present strategic climate-related matters.

Climate Risk Is Not Integrated with Readiness Measurement or Reporting

In addition to not being a central theme in readiness discussions, climate effects are not well integrated with readiness reporting and assessment. DoD and the services have robust and well-established processes for measuring, recording, and assessing readiness from the tactical level through the strategic level. The primary assessment processes—unit-readiness reporting and strategic-readiness assessment—are focused on short-term readiness and do not consider the potential readiness impacts of factors that may play out over the course of years or decades, such as climate change. Metrics are designed for the primary purpose of reporting upward about the current and short-term state of force readiness, failing to capture information about causes of disruptions and lacking the sensitivity needed to capture small perturbations and local adaptation activities that may ultimately accumulate to have a significant impact on readiness generation.

Current Processes, Tools, Models, and Assessments Show Promise as Points of Future Integration

While the findings above describe shortcomings of the current integration of climate risk and readiness decisionmaking and assessment, we found many potential ways that DoD and the services could use existing mechanisms to strengthen the connection between these two domains. The DoD Climate Adaptation Plan stresses the importance of actually "Implementing Climate-Informed Decision-Making," highlighting focus areas related to the incorporation of "Climate Intelligence" into all types of threat assessment, the incorporation of climate into "Strategic, Operational, and Tactical Decision-Making," and the incorporation of climate into "Business Enterprise Decision-Making."[5] Our literature review and outreach identify potential integration points between climate and readiness that speak to all three of these focus areas.

The Climate Readiness Framework Can Provide a Useful Structure for Understanding Readiness Vulnerability to Climate Risk

The framework provides a standardized means to organize the ways in which climate change could affect readiness and to subsequently visualize and analyze patterns among elements in the framework. These patterns can reveal areas of risk that may then be targeted for further investigation or prioritized for adaptive action, or both.

---

[5] U.S. Department of Defense, *Department of Defense Climate Adaptation Plan*, September 1, 2021b, p. 6.

The Full Impact of Climate Change on Readiness Requires Consideration of All Climate Hazard Pathways in Aggregate

Many climate impacts on readiness are currently seen as minor because of a perception that they may be temporally or spatially distant, easily mitigated through adaptation, or both. However, while minor hazards in isolation may be overcome relatively easily, a confluence of multiple pathways simultaneously could quickly aggregate to mission-level impacts in "death by a thousand cuts" fashion. Conversations that we had with stakeholders frequently included questions about the "most important" climate effect or the "biggest" hazard, but a review of the climate readiness framework and set of pathways shows that many hazards, through many mechanisms, could have similar impacts, leading to an aggregate effect on readiness inputs. Prioritizing these impacts and the kinds of disruptions that could occur through combined effects of multiple hazards and mechanisms may be more fruitful than attempting to prioritize the hazards themselves.

Patterns Among Pathways Can Be a First Step Toward Prioritization

Together, quantitative analysis of the frequency of and connections between elements of pathways and qualitative analysis of their relative sensitivity and adaptive capacity can inform preliminary decisionmaking regarding where and how to prioritize action to reduce or prevent impacts of climate change on readiness. The frequency of a hazard within the pathways should be understood as an indicator of the number of "opportunities" the hazard may have to affect readiness—and thus also as an indicator of the hazard's potential to disrupt readiness in multiple ways simultaneously, with possible amplifying effects—rather than as the sole determinant of priority. Similarly, focusing the frequency analysis on readiness inputs highlights those elements of the readiness system that have exposure to the widest variety of hazards and impacts. These initial indicators of a hazard's (or readiness input's) prevalence among the variety of pathways identified should form the basis for additional analysis to better understand the implications of exposure at specific geographic locations where the hazard or readiness input may be especially common or critical.

Information Beyond Hazard Representation in Pathways Is Required to Understand the Vulnerability of Readiness to Climate

Understanding a readiness input's vulnerability to climate and climate change requires consideration of the input's adaptive capacity and sensitivity to a variety of hazards. We performed a Delphi exercise to qualitatively assess the vulnerability of the various pathways to climate hazards. Overall, we found that the Delphi exercise is a useful, structured approach for generating a high-level characterization of the narratives or pathways and fostering a rich discussion. In particular, the approach highlighted key assumptions among participants about the drivers that influence a climate hazard's potential to affect a readiness input, as well as the available options and resources for adaptation. Results from this exercise suggest that inputs relating to facilities and large infrastructure and equipment investments may be the most vulnerable to future climate change, as these inputs are often tied to design standards and performance requirements that might not account for changing hazards and are often expensive to replace.

Contents

APPENDIXES

Available at www.rand.org/t/RRA1551-1

Climate and Readiness: Understanding Climate Vulnerability of U.S. Joint Force Readiness: Climate Hazard Pathways Appendix

Figures and Tables

Figures

Tables

CHAPTER 1

Introduction

The 2022 National Defense Strategy and the preceding *Interim National Security Strategic Guidance* note growing and accelerating global challenges to advancing and safeguarding vital U.S. national interests.[1] Increasing challenges stem not only from growing multidomain threats from strategic competitors but also from persistent threats from opposing nations and violent extremist organizations. These challenges are "not limited to man-made threats; DoD [the Department of Defense] must also execute its [mission-essential functions] in the face of disruptions caused by naturally occurring hazards and technological failures."[2] The Intergovernmental Panel on Climate Change's (IPCC's) Sixth Assessment concludes that global temperatures will undoubtedly increase at least through 2050 and, unless deep cuts to greenhouse gas emissions are made in the short term, will continue to rise beyond targets that were established to avoid severe or catastrophic consequences.[3] These changes will alter climate conditions and weather patterns globally. Such effects of climate change are occurring now, in every inhabited region, and are not constrained by national borders. Accordingly, the physical environment in which DoD must operate is being affected by climate hazards and is thereby adversely affecting the performance of the joint force and the systems that support it.

In March 2021, U.S. Secretary of Defense Lloyd J. Austin III issued a memorandum directing DoD "to include climate considerations as an essential element of our national security and to assess the impacts of climate change on our security strategies, operations, and infrastructure."[4] Generating, maintaining, and even increasing force readiness in light of changing climate threats is a key component of meeting the United States' highest-level strategic goals, from defending the homeland to deterring aggression and strategic attacks. Within DoD, *readiness* is defined as "the ability of military forces to fight and meet the demands of assigned missions."[5] Readiness is concerned with the ability to *organize, train, and equip* or *man, train, and equip* the force and is discussed at the *tactical, operational,* and *strategic* levels. *Climate* is the weather in a particular region averaged over a period, and *climate change* refers to long-term changes in the climate driven by global warming.[6] Although there exist both a significant body of literature on climate change impacts and many past studies on readiness and readiness risk, the Office of the Secretary of Defense (OSD) identified a need to establish both qualitative and quantitative links between climate effects and readiness impacts. This

---

[1] U.S. Department of Defense, *2022 National Defense Strategy of the United States of America*, October 27, 2022b; Joseph R. Biden, Jr., *Interim National Security Strategic Guidance*, The White House, March 2021b.

[2] U.S. Department of Defense, *Mission Assurance Strategy*, April 2012, p. 1.

[3] IPCC, "Summary for Policymakers," in Hans-Otto Pörtner, Debra C. Roberts, Melinda M. B. Tignor, Elvira Poloczszanka, Katja Mintenbeck, Andrés Alegría, Marlies Craig, Stefanie Langsdorf, Sina Löschke, Vincent Möller, et al., eds., *Climate Change 2022: Impacts, Adaptation and Vulnerability: Working Group II Contribution to the Sixth Assessment Report of the Intergovernmental Panel on Climate Change*, Cambridge University Press, 2022, p. 10.

[4] Lloyd J. Austin III, "Statement by Secretary of Defense Lloyd J. Austin III on Tackling the Climate Crisis at Home and Abroad," press release, U.S. Department of Defense, January 27, 2021.

[5] U.S. Department of Defense, *DOD Dictionary of Military and Associated Terms*, November 2021d, p. 179.

[6] U.S. Geological Survey, "What Is the Difference Between Weather and Climate Change?" webpage, undated.

report documents the results of an initial study aimed at developing such links between climate and readiness, laying the groundwork for the eventual integration of climate risk with quantitative readiness assessment and decisionmaking.

This study builds on previous RAND studies on installation resilience to natural hazards,[7] as well as on numerous efforts driven by DoD, the executive branch, and Congress to integrate climate change considerations within operational and infrastructure planning dating back to 2012, when the first "Climate Change Adaptation Roadmap" was issued by DoD in response to Executive Order 13514.[8] Since fiscal year (FY) 2019, the National Defense Authorization Act (NDAA) has outlined requirements to improve installation resilience to climate change, each building on requirements established in the previous year. In FY 2021, the NDAA required an update to DoD's 2014 Climate Adaptation Roadmap, which identified climate change–induced vulnerabilities for DoD in the areas of plans and operations (including readiness), training and testing of operations' capacities and safety, built and natural infrastructure, and weapon-system and equipment operational demands and logistics support. The FY 2022 NDAA further widened the aperture beyond installations to require assessments of climate impacts on operational missions, including analysis of impacts on force deployment, warfighting capability, and equipment resilience.[9]

In response to these requirements, and in line with President Joe Biden's Executive Order 14008,[10] DoD has produced a substantial number of recent reports, plans, tools, and strategy documents that assess and provide guidance for responding to climate-related threats. These include departmentwide "agenda-setting" documents, such as the DoD Climate Adaptation Plan and DoD Climate Risk Analysis, as well as numerous service- and region-specific documents, such as the Army Climate Strategy, Navy Climate Strategy, DoD Arctic Strategy, Navy Arctic Strategy, Air Force Arctic Strategy, and Army Arctic Strategy and a forthcoming update to the 2013 U.S. Arctic Strategy.[11] DoD has also been developing a climate exposure screening tool,

---

[7] Beth E. Lachman, Susan A. Resetar, Nidhi Kalra, Agnes Gereben Schaefer, and Aimee E. Curtright, *Water Management, Partnerships, Rights, and Market Trends: An Overview for Army Installation Managers*, RAND Corporation, RR-933-A, 2016; Beth E. Lachman, Anny Wong, and Susan A. Resetar, *The Thin Green Line: An Assessment of DoD's Readiness and Environmental Protection Initiative to Buffer Installation Encroachment*, RAND Corporation, MG-612-OSD, 2007; Anu Narayanan, Michael J. Lostumbo, Kristin Van Abel, Michael T. Wilson, Anna Jean Wirth, and Rahim Ali, *Grounded: An Enterprise-Wide Look at Department of the Air Force Installation Exposure to Natural Hazards: Implications for Infrastructure Investment Decisionmaking and Continuity of Operations Planning*, RAND Corporation, RR-A523-1, 2021.

[8] U.S. Department of Defense, *Department of Defense Climate Adaptation Plan*, September 1, 2021b; U.S. Department of Defense, *Climate Change Adaptation and Resilience*, DoD Directive 4715.21, January 14, 2016; Barack Obama, "Federal Leadership in Environmental, Energy, and Economic Performance," Executive Order 13514, Executive Office of the President, October 5, 2009.

[9] Public Law 116-283, William M. (Mac) Thornberry National Defense Authorization Act for Fiscal Year 2021, January 1, 2021; Council on Strategic Risks and Center for Climate and Security, *Climate Change and the National Defense Authorization Act*, June 2022.

[10] Joseph R. Biden, Jr., "Tackling the Climate Crisis at Home and Abroad," Executive Order 14008, Executive Office of the President, February 1, 2021a.

[11] U.S. Department of Defense, 2021b; U.S. Department of Defense, *Department of Defense Climate Risk Analysis*, October 2021c; Department of the Army, Office of the Assistant Secretary of the Army for Installations, Energy and Environment, *United States Army Climate Strategy*, February 2022; Department of the Navy, Office of the Assistant Secretary of the Navy for Energy, Installations, and Environment, *Department of the Navy Climate Action 2030*, May 2022; U.S. Department of Defense, *Department of Defense Arctic Strategy*, Office of the Under Secretary of Defense for Policy, June 2019; Department of the Navy, *A Blue Arctic: A Strategic Blueprint for the Arctic*, January 5, 2021; Department of the Air Force (DAF), *The Department of the Air Force Arctic Strategy*, July 21, 2020a; Department of the Army, *Regaining Arctic Dominance*, Chief of Staff Paper No. 3, January 19, 2021; Hilde-Gunn Bye, "The USA Will Release a New National Arctic Strategy," *High North News*, last updated April 8, 2022.

the DoD Climate Assessment Tool (DCAT), which builds on an existing tool developed by the U.S. Army Corps of Engineers.

Despite these enhanced efforts to situate climate change within a broad DoD agenda, numerous challenges complicate the incorporation of climate change effects into decisionmaking. While decisionmaking processes tend to focus on the short term and assume that historical weather patterns are unchanging, the full effects of climate change lag the accumulation of greenhouse gases into the atmosphere, with uncertain timing and magnitude.[12] Incorporating projections of climate effects into decisionmaking requires individuals or organizations to consider both slower-moving stressors and shorter-term events or shocks, as well as account for possibly much larger, less certain, and even cascading effects into the future.[13]

Establishing Common Vocabulary

Recognizing that this report covers a topic at the intersection of two individually vast disciplines, we offer some basic definitions of key terms that may be unfamiliar to readers coming to this work from one or the other area of expertise. Audiences who have a basic familiarity with climate and readiness terms may wish to skip this review.

Climate Science Vocabulary

Climate science and impact studies commonly use terms whose definitions may not align with those of similar terms used in the readiness community. Here, we define terms used throughout this report, to minimize ambiguity.

Weather refers to short-term atmospheric conditions. *Climate* is the weather in a particular region averaged over a period, typically on the order of decades to centuries. *Climate change* refers to long-term changes in the climate driven by global warming, which may manifest in a variety of phenomena, such as altered precipitation patterns and sea-level rise.[14]

Climate hazards are climate-driven physical events or trends that may cause loss of life, injury, or other health impacts, as well as damage to property, infrastructure, services, and ecosystems.[15] The IPCC does not recommend *hazard* as a generic term for climatic events or trends, since they may not have adverse consequences for all elements of an affected system; instead, it recommends the term *climatic impact driver*.[16] However, because we focus exclusively on those impacts that could have negative effects on readiness, we use the term *hazard* in this report.

---

[12] IPCC, "Summary for Policymakers," in Valérie Masson-Delmotte, Panmao Zhai, Anna Pirani, Sarah L. Connors, C. Péan, Sophie Berger, N. Caud, Y. Chen, Leah Goldfarb, Melissa I. Gomis, et al., eds., *Climate Change 2021: The Physical Science Basis: Contribution of Working Group I to the Sixth Assessment Report of the Intergovernmental Panel on Climate Change*, Cambridge University Press, 2021.

[13] Robert Lempert, Jeffrey Arnold, Roger Pulwarty, Kate Gordon, Katherine Greig, Cat Hawkins Hoffman, Dale Sands, and Caitlin Werrell, "Reducing Risks Through Adaptation Actions," in David Reidmiller, Christopher W. Avery, David R. Easterling, Kenneth E. Kunkel, Kristin Lewis, Thomas K. Maycock, and Brooke C. Stewart, eds., *Fourth National Climate Assessment: Vol. II, Impacts, Risks, and Adaptation in the United States*, U.S. Global Change Research Program, 2018.

[14] U.S. Geological Survey, undated.

[15] IPCC, 2022.

[16] Andy Reisinger, Mathias Garschagen, Katharine J. Mach, Minal Pathak, Elvira Poloczanska, Maarten van Aalst, Alexander C. Ruane, Mark Howden, Margot Hurlbert, Katja Mintenbeck, et al., *The Concept of Risk in the IPCC Sixth Assessment Report: A Summary of Cross-Working Group Discussions*, Intergovernmental Panel on Climate Change, September 4, 2020.

Climate *mitigation* refers to the process of taking action to reduce emissions of anthropogenic greenhouse gases to limit the long-term impact of climate change. This is distinct from *adaptation*, which refers to the process of adjustment to actual or expected climate and its effects in order to moderate harm.[17] Adaptation is closely related to the concept of *adaptive capacity*, or the ability to moderate harmful impacts through adaptation.[18]

Exposure generally refers to the presence of people, natural or built environments, or other phenomena (e.g., services; resources; and other economic, social, and cultural assets) in places and settings that could be adversely affected by a climate hazard.[19] More specifically, we define *exposure* as occurring when a climate hazard affects a particular input to readiness. The level of exposure can change over time.

Resilience refers to the ability to maintain essential function, identity, and structure.[20] It is often understood as a form of short-term adaptive capacity, describing an ability to bounce back from disruptions. DoD Directive 4715.21 defines *resilience* as the "[a]bility to anticipate, prepare for, and adapt to changing conditions and withstand, respond to, and recover rapidly from disruptions."[21]

Climate *sensitivity* is generally understood as susceptibility to harm. A population or system with high sensitivity and adaptive capacity may experience harm immediately following exposure but may be able to effectively moderate harm from future exposure in the longer term. Sensitivity and adaptive capacity are both elements of *vulnerability*, or the propensity or disposition to be adversely affected by climate change.[22]

Military Readiness Vocabulary

The military readiness community has developed specific, standardized language that may not be familiar to all readers. As stated above, DoD defines *readiness* as "the ability of military forces to fight and meet the demands of assigned missions,"[23] where *assigned missions* refers to all operational activities that a part of the armed force might be called upon to do by its commanders. Readiness is historically discussed in terms of the ability to *organize, train, and equip* or *man, train, and equip* the force. Accordingly, the concept of readiness can be applied to individuals, units, and equipment:[24]

- *Individual readiness* refers to the readiness of individual warfighters, including medical readiness, completion of necessary individual training and professional development, and mastery of specific skills.

---

[17] IPCC, 2022.

[18] These definitions vary slightly from the definitions used in the DoD Climate Adaptation Plan, which gives the following definitions:

> **Adaptation:** Adjustment in natural or human systems in anticipation of or response to a changing environment in a way that effectively uses beneficial opportunities or reduces negative efforts.

> **Resilience:** The ability to anticipate, prepare for, and adapt to changing conditions and withstand, respond to, and recover rapidly from disruptions. . . .

> **Mitigation:** Measures to reduce the amount and speed of future climate change by reducing emissions of heat-trapping gases or removing carbon dioxide from the atmosphere. (U.S. Department of Defense, 2021b, p. 2, emphasis in original)

[19] IPCC, 2022.

[20] IPCC, 2022.

[21] U.S. Department of Defense, 2016, p. 11.

[22] IPCC, 2022.

[23] U.S. Department of Defense, 2021d, p. 179.

[24] G. James Herrera, *The Fundamentals of Military Readiness*, Congressional Research Service, R46559, October 2, 2020, pp. 9–13, 39.

- *Training readiness* considers whether necessary training and education have been completed at various levels of the force, including individual, small unit, and larger unit (collective) training.
- *Equipment readiness* refers to the readiness of the equipment that enables units to achieve their missions, including maintenance, spare parts availability, and equipment condition.

Collectively, individual readiness, training readiness, and equipment readiness generate *unit readiness*, which refers to the ability of a unit to achieve its assigned mission set. Unit readiness can be considered at various levels, from large, complex units to the individual teams that compose them.

Much of the language related to readiness and readiness reporting is codified in Title 10 of the U.S. Code, which assigns the chairman of the Joint Chiefs of Staff the responsibility of "evaluating the overall preparedness of the joint force to perform the responsibilities of that force under national defense strategies and to respond to significant contingencies worldwide," subject to the authority, direction, and control of the President and the Secretary of Defense.[25] To carry out this responsibility, the chairman of the Joint Chiefs of Staff uses the Chairman's Readiness System. The Chairman's Readiness System considers joint readiness at three conceptual levels—strategic, operational, and tactical:

- *Strategic readiness* focuses on the overall ability of the joint force to perform the range of missions and to achieve the objectives outlined in strategic-level documents (such as the National Defense Strategy).
- *Operational readiness* focuses on the ability of combatant commanders to utilize ready joint forces to execute their specific assigned missions (such as numbered operational plans).
- *Tactical readiness* focuses on unit-level readiness, which is measured against a unit's ability to execute its designed or assigned mission, and takes into account a combination of metrics focused on resourcing (i.e., people and equipment) and training.[26]

Contribution of This Report

As mentioned above, this report documents the results of an initial examination of links between climate-driven hazards and readiness impacts, presenting both a framework for assessing these links and an assessment of the vulnerability of readiness to climate threats. More specifically, we introduce and populate a framework for enumerating, organizing, and assessing ways in which climate risk may affect global joint force readiness in the context of the current defense strategy and threat environment. The focus of this framework is the identification of possible disruptions to DoD's readiness-generation enterprise, meaning disruptions to the processes that yield ready warfighters, ready equipment, and ready formations that can be deployed and employed as needed. While weather already disrupts such processes in the short term, climate change may lead to different sets of weather effects on future readiness and have implications for strategic-readiness planning and future operational and tactical execution of readiness-generation processes. To reduce the study's scope so that it was commensurate with available time and resources, we did not attempt to cover climate impacts on the global threat environment or demand for DoD forces for missions directly related to climate- or weather-induced crises over a longer time horizon. However, we acknowledge that, even in the short term, significant readiness impacts may be driven by demand for "climate response" missions, such as humanitarian assistance and disaster relief (HADR) or Defense Support of Civil Authorities (DSCA) opera-

---

[25] 10 U.S.C. Section 153(a)(4)(A).

[26] Chairman of the Joint Chiefs of Staff, *CJCS Guide to the Chairman's Readiness System*, CJCS Guide 3401D, November 15, 2010, pp. 1–2.

tions responding to wildfires, floods, and other climate-related disasters. During such events, forces might be called on to assist civilian authorities with search and rescue and other disaster relief activities, rendering these forces unavailable for deployment elsewhere and potentially exposing them to unsafe conditions. Such events and the operational demands they generate constitute one of the most important climate-related impacts on readiness today, and they are likely to increase with future climate change. As described in Chapter 3, our framework provides a means for linking impacts from changing demand for DoD capabilities to readiness inputs, such as force projection, in order to identify mechanisms through which "the ability to provide capabilities required by the Combatant Commander to execute assigned missions" might be degraded.[27] Additionally, the framework provides a readiness-impact category that would allow future studies to explore changes to DoD's demand signals that are caused directly by climate. Because of the time and resources available to conduct this study, as well as the focus of the sponsor's interests, changes to DoD mission sets and the global security environment were specifically outside the scope of the study.

Our study stopped short of developing quantitative models or predictive analytic capabilities that will forecast climate impacts on readiness. However, the proposed framework provides a structure for understanding how exposure of readiness inputs to climate-related hazards may intersect with the underlying vulnerability of those inputs, potentially resulting in degraded DoD capabilities. Furthermore, the framework elucidates how the vulnerability of joint force readiness inputs stems from the combination of the sensitivity of those inputs to climate hazards and the capacity of DoD to moderate harmful impacts to readiness through adaptation.

Overview of This Report

The rest of this report outlines the methods and results of our research, as well as a set of recommendations for consideration by DoD senior leaders and other stakeholders. Chapter 2 documents information we collected through an extensive literature review and a broad set of conversations with DoD experts and others at the intersection of climate and national security. Chapter 3 introduces the *climate readiness framework*, which is based on this background research. Chapter 4 documents and describes the specific *climate hazard pathways* that populate the framework, providing an initial assessment of the vulnerability of joint force readiness to climate risk. Chapter 5 speaks to ways that the climate readiness framework may be applied to inform current and future integration of climate risk with readiness assessment, analytics, and decisionmaking, laying out a set of example *climate readiness integration points*. Finally, Chapter 6 summarizes study findings and recommendations.

---

[27] Chairman of the Joint Chiefs of Staff, 2010, p. 2.

CHAPTER 2

Current Understanding of Climate Impacts on Readiness

In this chapter, we offer a summary of the current state of integration between efforts to understand climate change and readiness in DoD. The trends and information described here are the result of a comprehensive literature review of both DoD and external documents, as well as a set of conversations with department, service, and external experts focused on both climate change and readiness. In the next sections, we give a brief overview of methods for the literature review and discussions, then summarize the general themes and trends that emerged from our engagements with a variety of readiness professionals.

Data Collection Methodology

To begin collecting information on the current state of thinking at the intersection of climate and readiness within DoD, we conducted a targeted literature review. Materials included DoD and service-specific strategy, policy, and guidance documents, as well as news articles, academic articles, assessment reports, and seminars on climate and security (U.S. and international).[1] We placed less emphasis on private-sector literature that could provide commercial analogues to DoD applications (e.g., general literature on equipment types, infrastructure and facilities, or people and their communities).[2] We also attended presentations and events that happened to occur during the study's timeline, including events hosted by the Center for Climate and Security and the American Security Project.[3] The goals of the literature review were to (1) uncover broad themes and trends related to the current state of climate thinking across the DoD readiness enterprise, (2) collect specific examples of how climate risk may affect readiness, and (3) identify additional sources, groups, and experts that could provide additional information on these topics.

Building on the literature review, we conducted semistructured virtual discussions with senior leaders, military officers, civilian officials, and subject-matter experts across various service and joint organizations who were knowledgeable about readiness issues, climate effects, and associated resourcing and deci-

---

[1] References used to populate the climate readiness framework (see, e.g., Chapters 3 and 4) are provided in a separate online appendix to this report (Katharina Ley Best, Scott R. Stephenson, Susan A. Resetar, Paul W. Mayberry, Emmi Yonekura, Rahim Ali, Joshua Klimas, Stephanie Stewart, Jessica Arana, Inez Khan, and Vanessa Wolf, *Climate and Readiness: Understanding Climate Vulnerability of U.S. Joint Force Readiness: Climate Hazard Pathways Appendix*, RAND Corporation, RR-A1551-2, 2023).

[2] Center for Climate and Security, homepage, undated; American Security Project, homepage, undated.

[3] Specific events attended include the Center for Climate and Security's "A Growing Crisis: The Launch of the World Climate and Security Report 2021" (June 7, 2021) and "Balancing on a Knife's Edge: Climate Security Implications of the IPCC Findings" (August 9, 2021), as well as the American Security Project's "How the United Kingdom Is Decarbonizing Defense & Adapting to Climate Change" (February 1, 2022).

sionmaking forums.[4] We held discussions with experts in the areas of weather, climate change effects on security, mission assurance (MA) planning, and military readiness. While most experts came from within DoD and the services, we also included selected external experts to gain a diversity of perspectives. The discussions included policy and resource groups (e.g., headquarters readiness, resource, and analytic offices); operational-readiness, training, and sustainment viewpoints (e.g., forces command readiness and future-plans offices); combatant commands (e.g., readiness and future-plans offices); international partners (e.g., North Atlantic Treaty Organization allies and partner nations in combatant commands outside Europe and North America); and areas of functional expertise (e.g., MA assessments, meteorological predictions, innovation and change analytics). Overall, we conducted approximately 30 engagements. Table 2.1 details the diversity of organizational perspectives that we captured.

We provided participants with background materials on the study's purpose, objective, and general focus areas for discussion. The stated intent was to examine general climate effects on service and joint readiness, with a focus on framing decisionmaking processes through appreciation of data-driven analytic capabilities, mission-related risks, and service strategies for climate adaptations. This intent allowed the participants to examine both general trends and observations, informing the material in this chapter, and more-specific examples, informing the development of the framework and pathways introduced in Chapters 3 and 4. As part of the introduction to each conversation, we explicitly stated that our discussion would focus not on how military missions and the global security environment might change, but rather on how climate might affect the ability to generate readiness for the kinds of missions DoD sees in the short-term environment.

The read-ahead materials for all discussions were intended to allow all individuals to properly prepare for the conversations and to include other thought leaders in the dialogue. Specifically, we stated that our discussions would be concentrated primarily on six focus areas:

- describing selected climate effects that, over time, could adversely affect service or joint readiness or increase mission risk such that DoD should preemptively adapt or invest to overcome unfavorable outcomes at a reasonable cost or limited loss of capabilities
- collecting information on past events that may have similarly demonstrated some effects of accumulated readiness impacts
- determining critical data elements needed by decisionmakers to enhance services' abilities to monitor, design, plan, invest, and adopt effective programs or adaptations across existing service processes
- assessing the accumulation of climate effects on readiness and exploring the determination of thresholds in transitioning from inconvenient friction to major catastrophe
- understanding and complementing existing or past service efforts relating to climate and readiness
- seeking other experts and representatives knowledgeable in these respective issues.

The information gathered through the literature review and expert engagements is the primary source for the general findings outlined in the rest of this chapter. As we will discuss in the following chapters, the same literature and conversations provided the starting point for development of the climate readiness framework and its component pathways, which lay out specific ways in which climate may affect readiness.

---

[4] The study was screened for the involvement of human subjects to ensure that their protection requirements were observed. Through an examination of the study's approach, methods, research protocols, and related materials, an independent body determined that the study was not a human subjects research project. That said, study participation, either full or in part, was voluntary, and no attribution is made to individual participants, although general references are made to the participants' respective organizations.

TABLE 2.1

Organizational Participation for Stakeholder Engagements

Organization	Unit
Military services	
Army	• Headquarters, Department of the Army/G-3/5/7, Readiness Division • U.S. Army Forces Command G3/Training and G5/Plans • Army Climate Change Working Group • Strategic-readiness tenet leads (x7)
Navy	• Naval Readiness Reporting Enterprise • Office of the Assistant Secretary of the Navy for Energy, Installations and Environment • Navy and Marine Corps Public Health Center • Navy Climate Working Group
Air Force	• Deputy Chief and Panel Chair, Readiness, Air Force/A3TR • Air Combat Command/Commander, 557th Weather Wing • Climate and Hydrology Plans, Headquarters Air Force/A3WX
Marine Corps	• Readiness Branch, Operations Division (PP&O/POR) • Critical Infrastructure and MA Program (PP&O/PS) • Navy and Marine Corps Public Health Center
Space Force	• Chief, Space Force Readiness Branch
Army National Guard	• Army National Guard G-9, Energy Branch
Combatant commands and service component commands	
USINDOPACOM	• Senior meteorology and oceanography officer • Climate Change Impacts Program Manager, Center of Excellence in Disaster Management
PACAF	• PACAF, Chief, Readiness and Integration Division • PACAF, Chief, Weather Operations Branch, A318
USTRANSCOM, SDDC	• Office of the Deputy Director, Transportation Engineering Agency
Others	
OSD(P&R)	• Service readiness liaisons
OSD(Policy)	• Office of the Under Secretary of Defense for Policy
UK MoD Climate Change and Sustainability Review	• Review lead
OSD(E&ER)	• Office of the Deputy Assistant Secretary of Defense for Environment and Energy Resilience

NOTE: E&ER = Environment and Energy Resilience; P&R = Personnel and Readiness; PACAF = Pacific Air Forces; PP&O = Plans, Policies & Operations; SDDC = Surface Deployment and Distribution Command; USINDOPACOM = U.S. Indo-Pacific Command; USTRANSCOM = U.S. Transportation Command.

Climate Risk Awareness in the Department of Defense and the Services

A Recent History

The seminal report on climate and national security is a 2007 CNA report that states that climate change has the potential to disrupt the United States, can exacerbate disasters, and functions as a threat multiplier in many regions of the world.[5] Shortly thereafter, a 2011 Defense Science Board study also linked climate change

---

[5] CNA Corporation, *National Security and the Threat of Climate Change*, 2007.

to national security concerns.[6] As far back as 2009, the Navy and the Army had working groups to address climate issues.[7] Since 2010, DoD has published sustainability performance plans that include statistics for energy and water use efficiency and greenhouse gas emissions. For example, the *Strategic Sustainability Performance Plan for Fiscal Year 2010* discusses maintaining readiness in the face of climate change and identifies ongoing activities that include vulnerability assessments of installation facilities and training ranges; inclusion of climate change effects in integrated natural resource management plans; and Strategic Environmental Research and Development Program funding of research into climate change assessment tools, the effects of sea-level rise on installation infrastructure, approaches to ecosystem management, and microgrid technologies.[8] As mentioned in Chapter 1, in 2014, DoD issued an update to its Climate Change Adaptation Roadmap, which identified climate change as a national security threat and detailed numerous vulnerabilities to climate change.[9] In a 2015 report to Congress, climate change was identified as a stressor in combatant commands' theater campaign plans for the United States, Africa, the Middle East, the Arctic, and South America. For example, U.S. Africa Command noted that climate change effects act as a threat multiplier that can increase the demand for humanitarian assistance and incorporated these effects into its annual Theater Campaign Plan reviews and partner nation engagement planning. USINDOPACOM planned for the consequences of sea-level rise in its concepts of operations for DSCA for pandemics and infection disease response and for HADR demands.[10] DoD has since issued additional road maps, guidance documents, and directives for integrating climate risk management into DoD planning processes.[11]

Climate-Related Groups

In early 2021, Secretary of Defense Lloyd J. Austin III established a climate working group, chaired by his special assistant, to coordinate DoD's climate and energy planning and to track the implementation of climate and energy initiatives.[12] Members include the under secretaries of defense for policy, research and engineering, acquisition and sustainment, and comptroller; the secretaries of the military departments; the chairman of the Joint Chiefs of Staff; the chief of the National Guard Bureau; and the director of the Office of Cost Assessment and Program Evaluation.[13] The group coordinated the development of the climate risk assessment, the DoD Climate Adaptation Plan, and associated annual updates, as required by Executive Order 14008.[14] The plan identifies lines of effort, explicitly tied to readiness, that are focused on increasing the resilience of built

---

[6] Defense Science Board, *Report of the Defense Science Board Task Force on Trends and Implications of Climate Change for National and International Security*, Office of the Under Secretary of Defense for Acquisition, Technology, and Logistics, U.S. Department of Defense, October 2011.

[7] Susan A. Resetar and Neil Berg, *An Initial Look at DoD's Activities Toward Climate Change Resiliency: An Annotated Bibliography*, RAND Corporation, WR-1140-AF, 2016.

[8] U.S. Department of Defense, *Strategic Sustainability Performance Plan for Fiscal Year 2010*, August 26, 2010.

[9] U.S. Department of Defense, *2014 Climate Change Adaptation Roadmap*, Office of the Assistant Secretary of Defense (Energy, Installations & Environment), June 2014.

[10] Resetar and Berg, 2016.

[11] Holly Kaufman and Sherri Goodman, *Climate Change in the U.S. National Security Strategy: History and Recommendations*, Council on Strategic Risks and Center for Climate and Security, Briefer No. 1, June 29, 2021; Resetar and Berg, 2016; U.S. Department of Defense, 2021b.

[12] David Vergun, "Action Team Leads DOD Efforts to Adapt to Climate Change Effects," DoD News, April 22, 2021.

[13] Secretary of Defense, "Establishment of the Climate Working Group," memorandum for senior Pentagon leadership, commanders of the combatant commands, and defense agency and DoD field activity directors, March 9, 2021.

[14] U.S. Department of Defense, 2021b; U.S. Department of Defense, 2021c.

and natural infrastructure and training and equipping a climate-ready force, with such listed enablers as continuously monitoring data analytics, incentivizing innovation, ensuring climate literacy, and pursuing environmental justice. The Office of the Deputy Assistant Secretary of Defense (E&ER) also has a climate action team to coordinate the office's efforts across installations, energy, operational energy, and climate resilience.[15] Although these groups generally represent only occasional commitments from senior officials, the existence and increased prevalence of such groups signals that climate is, increasingly, a DoD priority.

The departments of the Air Force, Army, and Navy also have working groups focused on climate- and energy-related issues. For example, the DAF has integrated climate threats into mission threat analyses and has developed guides or playbooks for installation self-assessment of threats, including natural hazards, and for mission sustainment. The Army issued its climate strategy and action plan and developed a guidebook for installation resilience investments.[16] The Department of the Navy is focused on the effects of climate on the people, training, and equipment essential to meeting the mission objectives of fleet operations, and its All-Hazard Threat Assessment (AHTA) process includes weather-related hazards, though not long-term projections of climate or energy demands.[17] In 2017, Naval Facilities Engineering Command published a guidebook to help planners consider climate hazards to installation facilities and infrastructure.[18]

The Department of Defense Climate Assessment Tool

Moving from strategy documents and planning groups toward initial analytic capabilities in the space, DoD chose to expand an existing geospatial tool developed by the U.S. Army Corps of Engineers in FY 2019. The Army Climate Assessment Tool (DCAT) provided screening-level assessments of climate hazard exposure at Army installations.[19] The tool's projections of exposure were based on outputs from an ensemble of 32 climate models, under two scenarios of "higher" (Representative Concentration Pathway [RCP] 8.5) and "lower" (RCP 4.5) greenhouse gas emissions[20] and for two 30-year epochs of projected climate centered at 2050 and 2085.[21] DCAT built on its predecessor in several ways. Climate hazards for which exposure would be considered were expanded to include coastal flooding, riverine flooding, drought, energy demand, heat, historical extremes, land degradation, and wildfire. Assessments of exposure were expanded across all departments, including 1,055 installations in the continental United States (CONUS), Alaska, and Hawaii and 336 installations in the rest of the world.

DCAT allows planners to assess installation exposure in a variety of ways. Users may browse reports generated from the National Standard View data archive, which weights all hazards equally, or adjust hazard weights to create custom assessments. Installations may be ranked by their aggregate exposure to all climate hazards, via an index known as the *Weighted Ordered Weighted Average* (WOWA) score, to enable compari-

---

[15] Office of the Deputy Assistant Secretary of Defense for E&ER personnel, discussions with the authors, December 8, 2021.

[16] Department of the Army, Office of the Assistant Secretary of the Army for Installations, Energy and Environment, 2022; A. O. Pinson, K. D. White, S. A. Moore, S. D. Samuelson, B. A. Thames, P. S. O'Brien, C. A. Hiemstra, P. M. Loechl, and E. E. Ritchie, *Army Climate Resilience Handbook*, U.S. Army Corps of Engineers, 2020.

[17] However, there is some emerging work to include future impacts within the AHTA (Navy Climate Working Group, discussion with the authors, October 6, 2021).

[18] Naval Facilities Engineering Command, *Climate Change Installation Adaptation and Resilience*, January 2017.

[19] A. O. Pinson, K. D. White, E. E. Ritchie, H. M. Conners, and J. R. Arnold, *DOD Installation Exposure to Climate Change at Home and Abroad*, U.S. Army Corps of Engineers, April 2021.

[20] The RCPs are a series of trajectories of atmospheric greenhouse gas concentrations adopted by the IPCC, representing different potential climate futures.

[21] Pinson et al., 2020.

FIGURE 2.1

DCAT Visualization of Relative Exposure to All Hazards at Continental U.S., Alaskan, and Hawaiian Installations for Lower- and Higher-Emissions Scenarios

Relative Exposure Across All Hazards

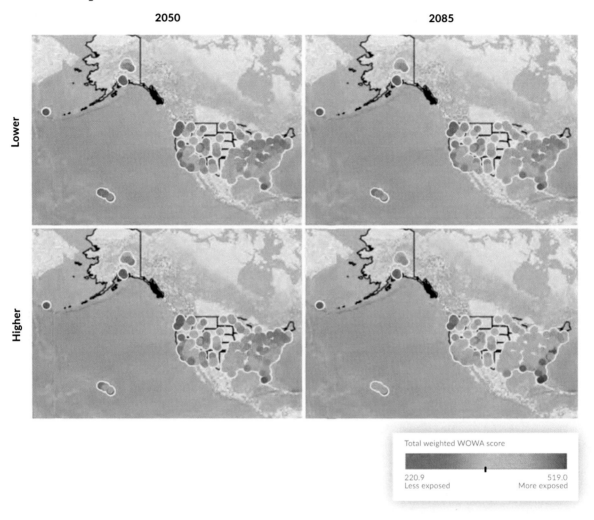

SOURCE: Adapted from Pinson et al., 2021.
NOTE: Lower-emissions scenarios are shown in the top half of the figure, and higher-emissions scenarios are shown in the bottom half. Data were pulled on February 28, 2021.

sons across military departments and regions (Figure 2.1). Individual installations may also be inspected in depth to identify the largest projected hazard exposure at each location and in each epoch. Users may investigate the relative contribution of specific hazard indicators to an installation's exposure score for a given hazard, as well as how those indicator values change across climate scenarios and epochs.

In its current form, DCAT is intended as a tool for DoD-wide hazard-exposure screening rather than for installation-level operational risk management or investment planning. It "[e]nables Military Departments and their installation personnel to deliver consistent exposure assessments and identify regions or installations for additional climate-related studies."[22] Additional analytic capabilities, such as assessments of hazard sensitivity and adaptive capacity, are in development.[23]

We learned in our discussions that, other than DCAT, which, as mentioned, is not suitable for local investment-level risk-management decisions, there is no authoritative data source for climate projections. Additionally, efforts to provide climate analytics capability, such as in Advana,[24] are in their infancy. The U.S. Air Force's 557th Weather Wing collaborates with the National Oceanic and Atmospheric Administration's National Centers for Environmental Information to produce weather reports with short-term forecasts or climate summaries, but they do not have climate modeling capabilities. Nor does a U.S. Navy detachment that is focused on weather and climate hazards have climate projection capabilities.[25] Therefore, MA planners and installation planners across the services must gather additional data locally to incorporate climate into their vulnerability analyses, plans, and assessments.[26]

Overall, we found that high-level goals laid out in DoD and service strategies are well articulated but are in early stages of implementation. Data and information availability, particularly related to climate projections, impacts, and resiliency investment opportunities, is not well established. We heard that, although the working groups are improving communication and coordination, there is a continued need for more interservice cross-fertilization.[27] Focus areas articulated in the DoD Climate Adaptation Plan are related to, for example, increasing installation or supply chain resilience to climate effects; few are focused on developing explicit links from climate effects to specific readiness inputs, determining priorities for investments to mitigate readiness risk, or understanding the cumulative effects on readiness.

Readiness-Specific Climate Risk Awareness

Climate Issues Generally Are Not Part of the Readiness Dialogue

Insights from our discussions indicated that although readiness issues may be a central theme in most discussions of climate effects, the converse is generally not true—climate effects, especially over the long term, are not part of the readiness dialogue or its reporting mechanisms. Discussion participants noted that threats to readiness levels are generally evaluated against short-term disruptors to the ability to execute specific missions; longer-term, broader considerations, such as climate, are somewhat of an afterthought. They also noted that the current fidelity of readiness measures, regardless of the level of assessment (from strategic to tactical), makes it difficult to discriminate what the respondents perceived to be relatively minor current effects of weather on readiness in the aggregate. (Generally, no one in the readiness or current operations communities has the longer perspective associated with broader climate issues.)

Discussion participants provided several reasons for these disconnects. First, the relevant time dimensions for climate and readiness are different, or at least appear different upon first inspection. Readiness leaders have immediate demands that often preclude them from dedicating attention to potential challenges that are above and beyond their efforts to accomplish closer tactical and operational goals. Even if short-term,

---

[22] U.S. Department of Defense, "DoD Climate Assessment Tool," fact sheet, April 5, 2021a.

[23] Government official, Office of the Deputy Assistant Secretary of Defense for E&ER, conversation with the authors, July 12, 2022.

[24] Advana is DoD's data analytics and big-data platform.

[25] Government personnel, OSD, discussions with the authors, October 5, 2021.

[26] Government personnel, Marine Corps and Army, discussions with the authors, January 19, 2022, and August 5, 2022.

[27] Office of the Deputy Assistant Secretary of Defense for E&ER personnel, discussion with the authors, December 8, 2021.

weather-related disruptions become more frequent, become more severe, or are otherwise affected by climate change trends, operational commanders tend to focus on the single event that must be adapted to rather than the effect of the over-time trend. Second, unit commanders typically deal with weather or climate issues as they arise. The relatively short duration of these short-term weather events is such that unit commanders may be able to plan their readiness activities for different locations, conduct evolutions at alternative times, sequence events differently, or incorporate appropriate simulations or simulators. Again, considering trends related to whether the incidence or nature of such disruptions is changing over time, driven by climate, is not the immediate focus of unit commanders. Third, weather- and climate-related issues tend not to be under a unit commander's direct areas of control or responsibility. Such broader and far-reaching climate responsibilities are apportioned to other, nonoperational leaders—installation commanders, regional commanders, and higher headquarters.[28] Alternatively, readiness reporting is specifically under the purview of a unit commander as they prepare and deploy their forces for assigned missions. Finally, if any planning has been conducted by units to address short-term weather issues, the actions typically involve only temporary, not long-term, actions and include such actions as asset dispersal, temporary relocation of required capabilities, or substitution of other capable units not affected by the respective weather events. Such decisions as permanent relocations, base closures, and shifting of certain operations to alternate locations are generally made by nonoperational leaders.

Climate Risk in Readiness Measurement and Assessment

Current Readiness Systems and Measures Reflect Units' Short-Term Status

Readiness professionals noted that the focus of the current readiness-assessment systems employed by the services and joint force is generally short term with respect to their measurement criteria for both available resources and unit capabilities. This orientation is especially true for the resource dimensions of people and training, although longer time horizons are already being considered for equipment and infrastructure for force generation and power projection. Readiness-planning time frames are commonly in months or, at a maximum, a few years, as the predictability of units' readiness significantly degrades because of the influence of external factors and a host of necessary assumptions. The influence of time on readiness assessments also varies considerably depending on the decisions being addressed: strategy determination, capability planning, resource allocation, operational planning, or risk mitigation. This is not to say that longer-term readiness perspectives are not considered or analyzed, but that such assessments tend to be more broadly based, highly aggregated, specifically caveated, and conditional on the decisions to be affected.

Other than the immediacy and observable impacts of specific short-term weather events (e.g., hurricanes, wildfires, tornados), the climate effects on readiness need to be assessed over a much longer time frame—several years, decades, or multiples of decades. In this manner, decisions about appropriate risk-reduction measures, investments, or other actions can be taken to avert or minimize effects on readiness. Like readi-

---

[28] Despite this informal designation of climate-related responsibilities by readiness leaders and commanders, installation commanders have been found to be lacking in executing public laws and DoD directives requiring them to address climate; energy risks; and threats to installation infrastructure, assets, and missions. Specifically, a DoD Inspector General report states that installation leaders at six Arctic and sub-Arctic locations "were unfamiliar with military installation resilience planning requirements, processes, and tools, and did not comply with requirements to identify current and projected environmental risks, vulnerabilities, and mitigation measures or incorporate these considerations into plans and operations" (Office of Inspector General, *Evaluation of the Department of Defense's Efforts to Address the Climate Resilience of U.S. Military Installations in the Arctic and Sub-Arctic*, U.S. Department of Defense, DODIG-2022-083, April 13, 2022, p. i). The report notes that "installation leaders focused on existing weather and energy challenges," were not provided with "guidance for implementing military installation resilience assessments," and "lacked resources to analyze and assess climate change" (p. i).

ness, climate model projections are typically regional to global in scale and become more uncertain as the time dimension increases. Therefore, while there may be similarities in terms of the broadness of model predictions, there is a distinct inconsistency between these two communities, as readiness is oriented toward a shorter time frame and climate effects reflect a longer time dimension.

Readiness Measures Lack Needed Sensitivity

We spoke to multiple readiness professionals who raised issues concerning the scope, level, and sensitivity of currently collected and systematically reported readiness information. This was especially true for the training dimension of readiness. Specifically, they noted that readiness training measures reflect only general training event participation and do not indicate the quality of training outcomes accomplished. Accordingly, these existing metrics provide limited differentiation among units or individuals who complete training evolutions. In the words of one participant, "The less we can do training exercises (because of climate inhibitors), the more we have to be able to measure the quality of training and the value of its resulting outcomes."[29] It follows that current readiness measurements—typically captured on a scale of 1 to 4—are lacking in the ability to distinguish between meaningful quantities of readiness. To address these deficiencies, either additional or more-detailed readiness information should be collected.

Interestingly, the U.S. Navy recently initiated a more in-depth process of examining its ship material casualty reports.[30] Although retrospective, this review of casualty reports is focused on root-cause analysis—i.e., what factors are systematically contributing to the degradation or loss of mission-area capabilities. The Navy's intentions are to examine all causal factors in readiness degradation. If climate issues are determined to be a significant and systematic contributor to readiness reductions, the Navy would elevate the issues within their respective resourcing or risk-mitigation processes to determine alternatives that could possibly counter the causal factors. Navy discussion participants identified a few issues that may emerge from a root-cause analysis: increased salinity adversely affecting the performance of undersea acoustics or accelerating material corrosion, and rising sea levels from ice melt and changing weather patterns affecting domestic and global ports so as to require greater resilience measures.

Finally, discussion participants noted that, if supplementary readiness data were to be collected for climate-relevant circumstances, additional guidance and policy specifications are needed. Such direction is needed to ensure consistency of information gathered across military services and thereby allow thorough comparisons and examination of trends. However, while participants agreed on the need for general policy direction, no individuals were willing to commit to the stringency of data specifications and certifications that were required for previous base realignment and closure activities.

Perceptions of Climate Impacts on Readiness

Perceived Climate Effects Are Locally Addressed, and Climate Risk-Mitigation Actions Are Not Tracked

Discussion participants informed us that unit commanders typically perceive any climate effects as a "training condition, and not as an operational threat." Accordingly, to the extent that such issues can be locally addressed, commanders implement necessary risk-mitigation actions to overcome the specific concerns

---

[29] Government official, Air Force A3-T, conversation with the authors, February 16, 2022.

[30] Casualty report ratings reflect critical issues with mission-essential equipment that cause either a major degradation or a loss of a mission area. Casualty reports are not filed if it is expected that the deficiency can be corrected within 48 hours.

or distractions, all of which are viewed as temporary or isolated issues. This distinct approach reflects the "can-do" attitude of commanders, and climate effects are thereby rarely reflected in readiness assessments.

The net result is that any actions taken by commanders to overcome climate issues are not recorded or accumulated over time. All risk-mitigation actions do incur some degree of "costs" (e.g., relocation to an alternative training range, loss in training fidelity, fewer or different personnel being qualified), but such details are not captured for holistic examination or analysis. Because this information is not systematically collected, senior leaders are unaware of the true costs of addressing climate-related issues and have limited insights into any progressive trends. Efforts to collect such data could seek to preclude a "death by a thousand cuts" situation or at least provide an early indication and warning if a particular location or activity is approaching a tipping point in transitioning from general climate annoyance to significant disruption.

Limited Lessons Are Carried Forward from Previous Events Affecting Readiness

Individuals with whom we spoke were unable to recall historical instances in which defense leaders had excelled at effectively documenting, measuring, analyzing, or communicating readiness impacts due to extraneous factors—instances that, had they been well documented, could have served as exemplars for assessing readiness impacts related to climate. To prompt the discussion, we highlighted several historical examples that could have provided possible insights: potential readiness impacts associated with adherence to environmental legislation (in 2004 to 2008),[31] continued annual congressional reporting on DoD's Readiness and Environmental Protection Integration Program,[32] consideration of military medical surveillance reports to examine trends in climate effects on personnel health and deployability,[33] substantiation of the readiness impacts associated with the Department of the Navy's closure of the Vieques training range,[34] and the relocation of military assets from and subsequent closure of Homestead Air Force Base due to the complete devastation resulting from Hurricane Andrew.[35] Each of these examples highlights experiences by DoD, or the affected service, that could have generated best practices or lessons learned that could have had implications for climate effects on readiness.

Multiple Factors Result in Units Operating Closer to the Margin

While the individuals with whom we spoke voiced a general disconnect between climate effects and readiness reporting, they consistently noted that military units are now "operating closer to the margin" as the cumulative impacts of readiness impediments are being realized. Such impacts include extended and sustained military deployments and operations, government sequestration, flatlined or declining defense budgets in key readiness areas, funding uncertainty associated with government continuing resolutions, cuts to equipment modernization, and lost training days due to local weather.

---

[31] U.S. Government Accountability Office, *Military Training: Compliance with Environmental Laws Affects Some Training Activities, but DoD Has Not Made a Sound Business Case for Additional Environmental Exemptions*, GAO-08-407, March 2008.

[32] A comprehensive repository of program details, primers, newsletters, congressional reports, and more is available at Readiness and Environmental Protection Integration, homepage, U.S. Department of Defense, undated.

[33] The *Medical Surveillance Monthly Report*, a peer-reviewed journal launched in 1995, provides monthly evidence-based estimates of the incidence, distribution, impact, and trends of health-related conditions among service members. See Health. mil, "Medical Surveillance Monthly Report," webpage, last updated 2022.

[34] U.S. House of Representatives, Implications of Closing the Vieques Training Facility: Hearing Before the Committee on Armed Services, June 27, 2001.

[35] Frederick J. Shaw, ed., *Locating Air Force Base Sites: History's Legacy*, Air Force History and Museums Program, 2004.

Discussion participants noted that, historically, the overall readiness-production function has had a degree of redundancy or a level of resilience that allowed system disruptions to overcome the uncertainty and unpredictability associated with generally "random" events or effects. Specifically, one readiness representative noted,

> Climate change is a factor that diminishes our trade space for making decisions. I always make trades with some degree of risk. As long as we continue to take nothing off the plate of items that we are responsible for, as the potential disruptors (like climate change) increase, our trade space decreases.
>
> As you continue to operate with the same missions, you are going to operate closer and closer to the threshold of failure. Climate change will continue to increase the readiness burden. The impact of climate change—the operational impact—is increased because we are operating at the margins.
>
> We no longer have enough headroom in our capacity and capability to respond to the disruptions.[36]

Integrating Climate Risk into Readiness Decisionmaking

Responsibility for Assessing and Addressing Climate-Readiness Issues Is Not Always Clear

Given that at least some participants in each discussion were generally readiness professionals, we observed a consistent theme around the topic of who has the overall responsibility or leadership for identifying and addressing climate-related readiness issues. This theme was consistent across the various ranges of seniority associated with the individuals with whom we spoke—from junior personnel (e.g., O4s–O5s, GS 12–GS 13) to more-senior leaders (e.g., O6s or general and flag officers, GS 15 and Senior Executive Service members). Most individuals noted that the issues were "above their pay grade" or generally required integrated approaches that extended beyond their direct areas of responsibility or resourcing. Potential adaptations or resource solutions can be substantial and can require considerable time for effects to be fully realized.

Strategic Foresight Needed to Influence Long-Term Decisionmaking Is Sometimes Lacking

The professionals with whom we spoke consistently noted a lack of strategic foresight and perspective regarding how climate issues are systematically considered across the variety of DoD decisionmaking forums. These individuals noted that DoD did not typically include climate factors in its broad deliberations (e.g., future operations and plans, resourcing, capability determinations and acquisitions, risk management, infrastructure investments, and climate as an intelligence factor), although such consideration is changing. The exclusion was attributed not necessarily to a lack of appreciation for weather or climate issues but rather to such functional expertise (1) not typically being present in decision forums, (2) not being tailored to present climate issues from a strategic and all-encompassing perspective, and (3) not having a dedicated organizational structure within which to systematically analyze and present strategic climate-related matters.

Involvement in Decisionmaking Forums

Professionals dedicated to military weather and climate readily admitted that they are not decisionmakers but acknowledge that it is important to have and to bring their science to bear in making high-level decisions. Such climate expertise—not necessarily uniquely but potentially as part of a cross-functional team—can

---

[36] Government official, Air Force A3-T, conversation with the authors, February 16, 2022.

inform and advise leaders about climate implications across such critical areas as planning, future operations, capability development, and acquisition programs. Professionals also noted that such inputs should occur throughout the full deliberation process—from problem specification to necessary data collection and modeling to development of alternatives to assessment of outcomes and refinements to the overall process.[37] Alternatively, several individuals identified the potential for weather and climate to be used to the strategic advantage of the U.S. military. This involves the climate community working closely with the intelligence community to assess how adversaries respond to weather conditions and to determine how these operational capabilities can be used to the advantage of the joint force from both an offensive perspective and a defensive perspective. In addition, with the continued development of space operations, the climate professionals should be involved to better understand, predict, and assess the implications of space climate and its potential impacts on critical decisions.

Strategic Education Involving Climate Expertise

Several climate professionals noted that DoD does not adequately prepare officers to think about climate effects. This educational deficiency is for both dedicated climate officers and operational officers in how best to use climate-related data and analyses. In addition, some individuals said that they did not see these circumstances necessarily improving: For example, one professional said, "I do not think the services will change their training and education approaches without a demand signal that indicates that there is either a significant capability requirement or gap." This attention to the need for strategic education involving climate expertise also includes access to, understanding of, analysis of, and decisionmaking with relevant databases and information sources. There are many disparate databases documenting weather-related impacts or potential impacts; however, several individuals referenced minimal efforts to aggregate this information, maintain a historical record, or assess for weather-related impacts. For example, such potentially relevant sources could include commanders' operational reports that are submitted when operations are being planned, started, stopped, and summarized. One individual noted that while such "data exists in some form, there is no centralized capability (or demand function) to analyze the data from a different perspective other than it[s] original or primary purpose." The individual noted that such an analytical perspective should be part of the strategically oriented education.

Dedicated and Long-Term Climate-Focused Organization

We conducted several discussions with USINDOPACOM's Climate Change Impacts (CCI) program. This one-of-a-kind program was established in June 2021 "to understand the threats, increase resilience, reduce fragility, and subsequently increase stability throughout the region."[38] CCI reports directly to the commander of USINDOPACOM and works collaboratively with civilian and military agencies in other Indo-Pacific nations that are responsible for climate security. Similarly, CCI interfaces with and seeks expertise from a variety of academic organizations, think tanks, nongovernmental organizations, and government

---

[37] One individual raised an example of his counterparts being involved in only the very final stages of an operational planning process for several quick-turn Korea scenarios that were determined to be infeasible because of short- and long-term weather conditions. Unfortunately, these insights were considered very late, such that considerable reworks were necessary and critical time was lost in the development of operational plans for a series of joint wargames.

[38] Center for Excellence in Disaster Management and Humanitarian Assistance, "Climate Change Impacts," webpage, undated.

security organizations to develop and undertake projects of mutual interest to enhance climate resilience in the region. Through these relationships, CCI does the following:

- Conduct research analyses and reports on climate change impacts on regional security.
- Facilitate integration of climate change impacts into INDOPACOM's regional exercises, training events and security cooperation programs.
- Establish and support an informal network of Indo-Pacific military and civilian national climate security officers—the Community for Indo-Pacific Climate Security (CIPCS)—to exchange information on national efforts and consider opportunities for international cooperation to enhance climate security.
- Conduct bilateral discussions with partner nations on areas of mutual concern and common interest.[39]

In our overall review, we were not able to find other dedicated organizations like CCI that are uniquely organized to address the challenging interactions of climate and military readiness. We acknowledge the need for and potential of such an organization but are not able to assess its contributions given its relatively short history. That being said, multidisciplinary efforts specifically oriented to understanding and assessing climate impacts from a regional perspective are warranted and should be further explored.

The Need for a Framework

In light of these findings related to the current state of integration between climate and readiness, we developed a framework and related products to help DoD begin to build additional connective tissue between the domains of climate science and joint force readiness. The rest of the report introduces the framework and its components, then speaks to how the framework could be used to inform a closer link going forward.

---

[39] Center for Excellence in Disaster Management and Humanitarian Assistance, undated.

CHAPTER 3

The Climate Readiness Framework

As discussed in this report's introduction, there is growing awareness that climate change is transforming the context in which DoD operates.[1] Climate-related changes have the potential to significantly affect joint force readiness. The IPCC has clearly stated that the world is experiencing the effects of a changing climate today.[2] A recent National Intelligence Estimate assessed the physical and geopolitical risks to readiness from climate change to be low in 2030, but they will rise to medium within the next ten years.[3]

To our knowledge, there has not yet been a systematic effort to capture, organize, or model potential climate change effects on joint force readiness. Rather, climate effects on readiness to date have been largely documented as anecdotes. As noted in Chapter 1, a key contribution of our study is a framework for understanding the risk to readiness that may result from a combination of climate hazard exposure and the underlying vulnerability of a variety of readiness inputs to that exposure. In keeping with the IPCC's Sixth Assessment report, *vulnerability of readiness inputs* is understood here to encompass such concepts as sensitivity and adaptive capacity.[4] In this chapter, we present the framework as an attempt to catalog the ways in which climate change can introduce risk into readiness processes in a way that can be readily extended for more-detailed or quantitative analysis.

Framework Structure

The *climate readiness framework* comprises three main sections, showing how *climate hazards*, which sometimes lead to causally important *second-order effects*, lead to *readiness impacts* that then affect relevant *readiness inputs*. This structure, shown in Figure 3.1, allows for standardization of language and application of a consistent organizational structure to linkages from climate, at the top, to readiness, at the bottom. Every path from a climate hazard (top) to a readiness input (bottom) tells the story of a particular way in which climate may affect joint force readiness. We call these stories *climate hazard pathways*. In some cases, one or more impacts on readiness may be caused by more than one hazard; for example, equipment damage and increased equipment maintenance requirements caused by freeze-thaw cycles may be caused by either extreme heat or extreme cold. We consider such cases as two separate pathways: one for each hazard, each resulting in one or more impacts on a single readiness input. Specific pathways are described in more detail in Chapter 4.

Placing climate hazard pathways into the readiness framework requires identifying the relevant climate hazards, readiness impacts, and readiness inputs using the standardized categories of the framework. While

---

[1] U.S. Department of Defense, 2022b.

[2] IPCC, 2022.

[3] National Intelligence Council, *National Intelligence Estimate: Climate Change and International Responses Increasing Challenges to US National Security Through 2040*, NIC-NIE-2021-10030-A, October 2021.

[4] IPCC, 2022.

FIGURE 3.1

The Climate Readiness Framework

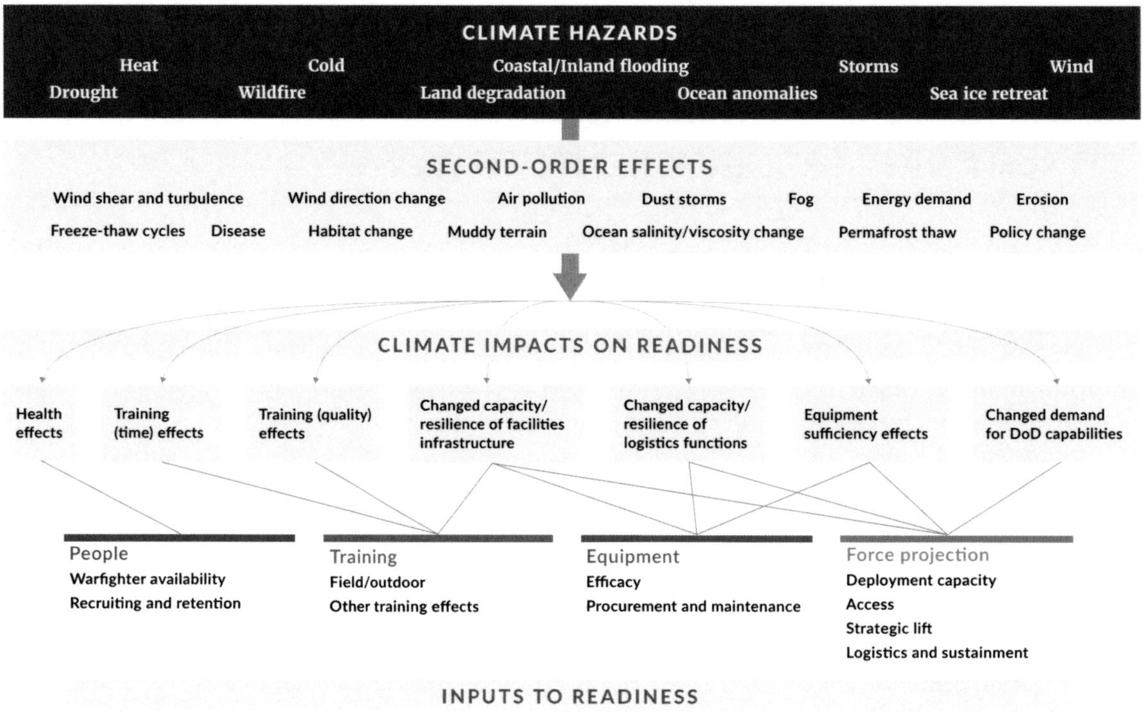

the items within each of these categories may depend on the scope of any specific application of the framework, and may shift over time (e.g., if additional climate hazards not enumerated here are found to affect readiness as climate change severity increases), the categories shown in the figure are intended to be complete for the purposes of this study's objective—that is, to better understand the ways in which climate risk may affect global joint force readiness in the medium term in the context of the current defense strategy and threat environment. The standardized language of the framework is useful because it facilitates the identification of patterns and specific points of interest among the pathways. In the following section, we introduce the standardized climate hazards, readiness impacts, and readiness inputs that make up the framework. The framework stops short of connecting climate impacts on readiness inputs to the downstream *consequences* that may result from the impacts; for example, one might use the framework to identify hazards that could delay training, but not the broader consequences of that delay. In the following sections, we introduce the three framework concepts in a different order from that of Figure 3.1: We first define the set of climate hazards, then define the four readiness inputs and their subcategories, and finally discuss the readiness impacts that connect the "top" and "bottom" of the framework.

Climate Hazards

The climate readiness framework identifies 13 climate hazards that may lead to impacts on readiness inputs. These hazards were selected based on our literature review and stakeholder inputs described in Chapter 2, and we attempted to align them as closely as possible to the hazards used in DCAT, most of which are designed to address installation vulnerabilities. Six of the eight hazards in DCAT are represented in our framework in some form (*drought, coastal flooding, inland flooding* [renamed from *riverine flood risk*], *heat, wildfire,* and *land degradation*), while a seventh hazard (*energy demand*) is included in the framework as a second-order effect. We separated the *historic weather extremes* hazard from DCAT into three hazards: *tropi-*

cal storms, winter/ice storms, and *wind*. Finally, drawing on our literature review and stakeholder outreach, we included four additional hazards: *cold, ocean anomalies, other precipitation* (i.e., precipitation unrelated to storm events), and *sea ice retreat*.

Drought

Drought refers to a deficiency of precipitation over an extended period that results in a water shortage.[5] Drought is found to be a possible disruptor of all four readiness inputs (people, training, equipment, and force projection).

Coastal Flooding

Coastal flooding is when water inundates or covers normally dry coastal land because of high or rising tides or storm surges.[6] Coastal flooding is found to be a possible disruptor of all four readiness inputs.

Inland Flooding

Inland flooding refers to an overflow of water onto normally dry land.[7] Inland flooding is found to be a possible disruptor of all four readiness inputs.

Heat

Heat refers to both an increase in overall average temperature and an increased frequency of extreme heat events. The definition of an *extreme heat event* varies geographically and with other contextual factors but is generally understood to be a period of at least two to three days with temperatures above 90°F and high humidity. Heat is found to be a possible disruptor of all four readiness inputs.

Wildfire

Wildfire refers to an unplanned fire burning in natural or woodland areas, such as forests, shrublands, grasslands, or prairies.[8] Wildfires are a significant threat to structures and communities within or near vegetated wildland areas, including many military bases, and they are made more likely by heat, drought, and high fuel loads in these wildland areas. Wildfire is found to be a potential disruptor to all four readiness inputs.

Land Degradation

Land degradation is characterized by the deterioration or loss of the productive capacity of the ecosystem. In the Arctic, land degradation is associated with soil warming and permafrost thaw, which causes subsidence and damage to infrastructure. Land degradation is associated with all four readiness inputs, although the majority of land degradation pathways are related to force projection.

Tropical Storms

Tropical storms refers to hazards of extreme wind and rainfall specifically associated with tropical storm events, including hurricanes. Tropical storms are subsumed within the hazard category *historic weather extremes* in DCAT but are treated separately from other extreme events in our framework. Tropical storms are found to be a possible disruptor of all four readiness inputs.

[5] Federal Emergency Management Agency, National Risk Index, "Drought," webpage, undated-b.

[6] Federal Emergency Management Agency, National Risk Index, "Coastal Flooding," webpage, undated-a.

[7] National Weather Service, "Flood and Flash Flood Definitions," webpage, undated.

[8] Federal Emergency Management Agency, National Risk Index, "Wildfire," webpage, undated-c.

Winter/Ice Storms

Winter/ice storms refers to hazards of extreme wind, cold, and frozen precipitation (i.e., snow, hail, and freezing rain) associated with winter storm events, such as ice storms and blizzards. Winter/ice storms are subsumed within the hazard category *historic weather extremes* in DCAT but are treated separately from other extreme events in our framework. Winter/ice storms are found to be a possible disruptor of the force projection and training readiness inputs.

Wind

Wind refers to wind effects unrelated to tropical or winter storms and includes high or heavy winds over land and water. Such effects are captured indirectly in DCAT as tornado effects under the hazard category *historic weather extremes*. We treat wind as a separate hazard in our framework to capture effects of wind that do not rise to the level of severity associated with extreme events. Wind is associated with potential disruptions to equipment and force projection readiness inputs.

Cold

Cold refers to climate events characterized by extreme cold, wind chill, and frost and/or freeze effects. These are periods of low temperatures (air and wind chill) that are at least equal to local or regional advisories (generally −18°F for air temperature and −35°F for wind chill), as well as air temperatures at or below 32°F or the presence of ice crystals on the ground that cause damage.[9] Cold is captured indirectly in DCAT in the hazard category *historic weather extremes* but is treated separately in our framework because of significant evidence that cold could disrupt the training, equipment, and force projection readiness inputs.

Ocean Anomalies

Ocean anomalies refers to changes in ocean salinity, viscosity, and/or carbon dioxide storage that ultimately stem from ocean warming and/or freshening due to land-based glacier or ice sheet meltwater runoff. This hazard is not included in DCAT, but we added it to our framework because of evidence that such systemic ocean changes could affect naval operations. Ocean anomalies are found to be a potential disruptor to the equipment readiness input.

Other Precipitation

Precipitation (other) refers to changing precipitation patterns unrelated to storms and drought. We added this hazard to the framework owing to evidence that long-term changes in precipitation, whether wetter or drier, may endanger animal and plant species and increase the sensitivity of natural environments, which may disrupt training activities.

Sea Ice Retreat

Sea ice retreat refers to long-term reduction in sea ice extent, primarily in the Arctic. This hazard is not included in DCAT, but we added it to the climate readiness framework owing to evidence of its possible effects on naval operations and its contribution to land degradation. Sea ice retreat is found to be a potential disruptor of the force projection readiness input.

Second-Order Effects

Second-order effects of climate hazards are included in the climate readiness framework as an additional means of categorizing climate hazards' pathways. Identifying a second-order effect for a pathway is not

---

[9] National Weather Service, *Storm Data Preparation*, National Weather Service Instruction 10-1605, July 26, 2021.

required, but it can help highlight important commonalities and link the climate hazard more directly to its impacts on readiness. More specifically, a second-order effect is a phenomenon that is caused by a climate hazard but is the more direct driver of the impact. For example, wildfires cause air pollution, which in turn can harm health, damage equipment, and make certain types of activities difficult, making pollution the proximal cause of the disruption to training or operations. The 14 second-order effects included in the current application of the framework are as follows, although additional second-order effects may be needed as more pathways are identified:

- air pollution
- disease
- dust storms
- energy demand
- erosion
- fog
- freeze-thaw cycles
- habitat change
- muddy terrain
- ocean salinity and/or viscosity change
- permafrost thaw
- policy change
- wind direction change
- wind shear and turbulence.

Readiness Inputs

The climate readiness framework includes four readiness inputs: *people, training, equipment,* and *force projection.* These inputs are rooted in the language currently used by DoD and the services to discuss readiness. They parallel the resources tracked by joint readiness reporting, namely personnel (P-ratings), equipment (S-ratings for availability and R-ratings for readiness), and training (T-ratings). We included force projection as an additional dimension owing to the fact that the Chairman's Readiness System defines *tactical readiness* as "the ability to provide capabilities required by the Combatant Commander to execute assigned missions."[10] Although tactical readiness focuses largely on unit readiness, we decided that including DoD's ability to project forces forward to the combatant commands was worthwhile. The four readiness inputs are essentially labels that can be applied to the climate hazard pathways, placing them into broad bins to allow examination of particular types of pathways but adding relatively little information on the consequences of the impacts on readiness that might occur.

To convey more information on what kind of disruption might occur within a pathway, we define subcategories associated with each of the four readiness inputs. As we describe in the next chapter, climate hazard pathways are enumerated in such a way that each pathway has exactly one associated readiness input, although it may have multiple associated readiness input subcategories and be driven by multiple climate hazards.

---

[10] Chairman of the Joint Chiefs of Staff, 2010, p. 2.

People

The *people* readiness input captures impacts to personnel readiness and individual people required for a ready force. This input has two subcategories:

- *Warfighter availability* refers to effects on individual warfighter availability, including deployability or ability to participate in training events, due to permanent or temporary physical or mental health effects for the individual warfighter.
- *Recruiting and retention* refers to effects on personnel readiness due to interruptions or changes to recruiting, accession, and retention of personnel with needed skill sets.

Training

The *training* readiness input captures impacts to training readiness, training events, and the ability to train. This input has two subcategories:

- *Field and outdoor training* refers to effects on training readiness due to changes in the environmental conditions at the location where training is conducted, such as changes in efficacy, realism, frequency, or required types of outdoor or field training activities, which may include the need to shift to simulated environments.
- *Other training* refers to effects on training readiness due to a change in something other than the (outdoor) environmental conditions at the location where training is conducted, including effects on indoor, classroom, or schoolhouse training.

Equipment

The *equipment* readiness input captures impacts to equipment readiness. This input has two subcategories:

- *Efficacy* refers to effects on equipment readiness due to degraded efficacy, usability, or applicability given a changing environment. This includes issues with equipment location and storage, such as Army Prepositioned Stock.
- *Procurement and maintenance* refers to effects on equipment readiness due to equipment procurement and maintenance issues, including degraded lifespan of equipment, additional maintenance requirements, or parts and platform procurement challenges. Global supply chain effects that disrupt the industrial base are also included here.

Force Projection

The *force projection* readiness input captures impacts to deployment readiness, access, and lift. This input has four subcategories:

- *Deployment capacity* refers to effects on deployment readiness due to disruption of transportation infrastructure, networks, and capacity, encompassing everything up to and including ports of embarkation and ports of debarkation. Climate impacts related to demand for DoD capabilities, such as units being unavailable for deployment because of HADR and DSCA disaster response activities, are linked to this subcategory.
- *Logistics and sustainment* refers to effects on deployment readiness due to disruption of operational logistics and sustainment issues for forward-deployed forces.
- *Strategic lift* refers to effects on deployment readiness due to disruptions in strategic lift (airlift and sealift from port of embarkation to port of debarkation).

- *Access* refers to effects on deployment readiness due to disrupted or changed access to forward locations and access rights.

It is important to note that the four readiness inputs used in the framework are highly interrelated. For example, the ability to conduct successful training events or effectively project force depends on having the correct set of people and equipment in place. Similarly, the ability to project capable forces requires having completed necessary training. The framework simplifies this complexity by identifying the single readiness input that is *most proximal* to the climate-related disruption. In other words, although in some cases a climate hazard starts a causal chain that crosses multiple inputs, the framework identifies only the first readiness input in this chain. For example, if personnel health is affected, a disruption to *training* might eventually occur; however, the most proximate readiness input would be identified as *people*.

Climate Impacts on Readiness

While the readiness-input subcategories begin to shed some light on *what* a climate hazard might disrupt, these categories do not yet capture the *mechanisms* by which climate disrupts readiness. To apply standardized language to these mechanisms, the climate readiness framework articulates a standardized set of impacts on readiness to classify the ways in which hazards may disrupt the inputs. While each pathway identifies a unique mechanism through which one or more hazards could affect a readiness input, the climate *impacts on readiness* provide some standard categories into which to place these mechanisms.

Seven categories of climate impacts on readiness are included in the climate readiness framework. These categories were developed from our attempt to identify commonalities among the various impacts on readiness that we collected from our literature review and stakeholder outreach. Six of the seven impacts generally correspond to phenomena germane to the process of generating readiness that might be disrupted. These impacts are as follows:

1. *Health effects:* impacts where one or more hazards cause permanent or temporary negative effects on physical health, mental health, or quality of life. Health effects are generally associated with the *people* readiness input.
2. *Equipment sufficiency effects:* impacts where one or more hazards make current equipment insufficient or more difficult or costly to employ, reduce equipment lifetime, cause damage, and/or lead to a need for increased maintenance. Effects that imply possible needed future changes in requirements are included here. Equipment sufficiency effects are generally associated with the *equipment* and *force projection* readiness inputs.
3. *Training (time) effects:* impacts where one or more hazards permanently or temporarily constrain training windows in terms of time of day or season. Training time effects are generally associated with the *training* readiness input.
4. *Training (quality) effects:* impacts where one or more hazards affect the type and/or quality of training that can be achieved, or the alignment of training to anticipated deployment conditions, including training realism, changing training environments, and related issues. Training quality effects are generally associated with the *training* readiness input.
5. *Changed capacity and resilience of facilities infrastructure:* impacts where one or more hazards disrupt, damage, or destroy facilities and related infrastructure. Changed capacity and resilience of facilities infrastructure is generally associated with the *training, force projection*, and *equipment* readiness inputs.
6. *Changed capacity and resilience of logistics functions:* impacts where one or more hazards lead to loss of, disruptions to, or uncertainty in supply chains, distribution networks, and other logistics func-

tions, including roads and rail lines. Changed capacity and resilience of logistics functions is generally associated with the *force projection* and *equipment* readiness inputs.

In addition to these impacts that are directly related to the readiness-generation process, we include a seventh impact:

7. *Changed demand for DoD capabilities:* impacts where one or more hazards lead to a change in level of demand for DoD capabilities, or a change in required force size, deployment frequency, or expected environment of current missions. Changed demand for DoD capabilities is generally associated with the force projection readiness input.

While changes to DoD mission sets and the global security environment were specifically outside the scope of our study, we found that the potential readiness impacts from such changes are significant enough that a framework that does not include this category would be incomplete and hard to adapt for other studies. We therefore include this impact in the framework's structure, with impacts on readiness inputs then being a result of these changes in demand. As we discuss in the next chapter, only two pathways in our study have this type of readiness impact: (1) a changed security environment due to climate and (2) increased demand for HADR and DSCA missions. The inclusion of these pathways simply serves to acknowledge the existence of these mechanisms for disruption to readiness without exploring this topic in a level of detail that is outside the purview of this study.

Figure 3.2 shows how climate impacts on readiness map to the four readiness inputs described above, providing a more detailed representation of the linkages illustrated in Figure 3.1. The set of connections between impacts and inputs shown in the two figures represents the results of our study's application of the framework and may change over time as additional pathways are identified. Figure 3.2 shows that three of the four readiness inputs have multiple types of climate impacts that could affect those inputs. The exception is *health effects*, which is associated with only the *people* input. Although health effects do ultimately affect, for example, training, the framework makes a simplifying assumption that identifies only *the most proximal* linkage between an impact and an input. The health impact on individual warfighters is the first step in any disruptive "causal chain" that starts with health effects. We apply the same assumption wherever a particular impact could implicate more than one readiness input. Where impacts truly do implicate multiple inputs independently (rather than sequentially in a causal chain), the framework draws multiple links; for example, the *changed capacity and resilience of facilities infrastructure* impact could affect training, equipment, and force projection, but the effect on one of these inputs does not occur as a result of the effect on one of the other two.

Possible Benefits of the Climate Readiness Framework

The climate readiness framework provides a standardized way to collect, catalog, and organize the ways in which climate change could affect joint force readiness. It can be thought of as an inventory of the potential ways in which different climate hazards can introduce risk to force readiness,[11] and it can help identify patterns and commonalities among the known climate hazard pathways. These patterns, in turn, can help users

---

[11] IPCC guidance characterizes risk from climate change as arising from effects that result from hazards acting on human systems and ecosystems. Risk is also introduced by responses to climate hazards. The focus in this study was on the ways in which climate hazards introduce risk to force readiness, not on the ways in which a potential response could introduce risk to readiness.

FIGURE 3.2

Climate Impacts on Readiness and Associated Readiness Inputs

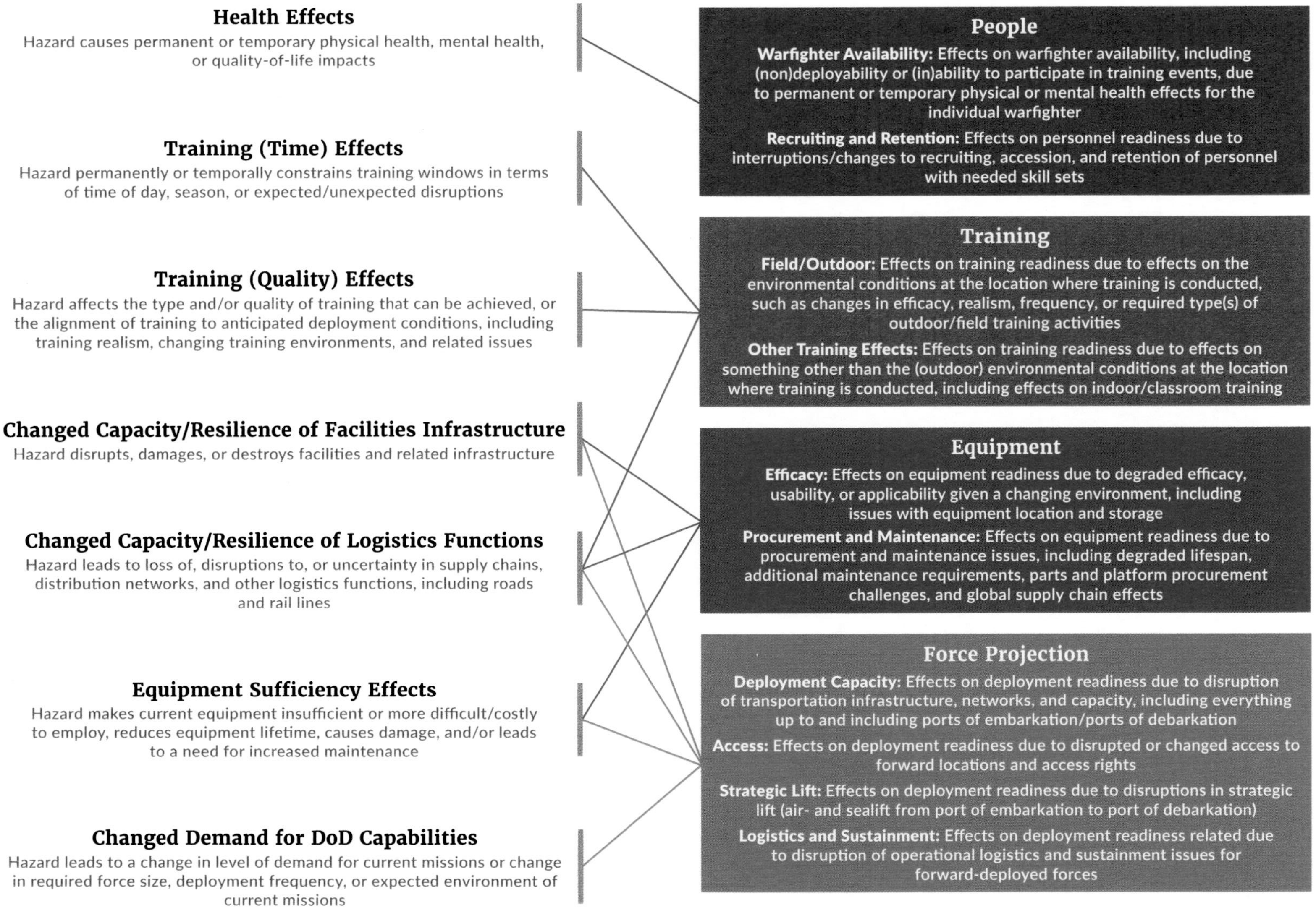

People

Warfighter Availability: Effects on warfighter availability, including (non)deployability or (in)ability to participate in training events, due to permanent or temporary physical or mental health effects for the individual warfighter

Recruiting and Retention: Effects on personnel readiness due to interruptions/changes to recruiting, accession, and retention of personnel with needed skill sets

Training

Field/Outdoor: Effects on training readiness due to effects on the environmental conditions at the location where training is conducted, such as changes in efficacy, realism, frequency, or required type(s) of outdoor/field training activities

Other Training Effects: Effects on training readiness due to effects on something other than the (outdoor) environmental conditions at the location where training is conducted, including effects on indoor/classroom training

Equipment

Efficacy: Effects on equipment readiness due to degraded efficacy, usability, or applicability given a changing environment, including issues with equipment location and storage

Procurement and Maintenance: Effects on equipment readiness due to procurement and maintenance issues, including degraded lifespan, additional maintenance requirements, parts and platform procurement challenges, and global supply chain effects

Force Projection

Deployment Capacity: Effects on deployment readiness due to disruption of transportation infrastructure, networks, and capacity, including everything up to and including ports of embarkation/ports of debarkation

Access: Effects on deployment readiness due to disrupted or changed access to forward locations and access rights

Strategic Lift: Effects on deployment readiness due to disruptions in strategic lift (air- and sealift from port of embarkation to port of debarkation)

Logistics and Sustainment: Effects on deployment readiness related due to disruption of operational logistics and sustainment issues for forward-deployed forces

Health Effects

Hazard causes permanent or temporary physical health, mental health, or quality-of-life impacts

Training (Time) Effects

Hazard permanently or temporally constrains training windows in terms of time of day, season, or expected/unexpected disruptions

Training (Quality) Effects

Hazard affects the type and/or quality of training that can be achieved, or the alignment of training to anticipated deployment conditions, including training realism, changing training environments, and related issues

Changed Capacity/Resilience of Facilities Infrastructure

Hazard disrupts, damages, or destroys facilities and related infrastructure

Changed Capacity/Resilience of Logistics Functions

Hazard leads to loss of, disruptions to, or uncertainty in supply chains, distribution networks, and other logistics functions, including roads and rail lines

Equipment Sufficiency Effects

Hazard makes current equipment insufficient or more difficult/costly to employ, reduces equipment lifetime, causes damage, and/or leads to a need for increased maintenance

Changed Demand for DoD Capabilities

Hazard leads to a change in level of demand for current missions or change in required force size, deployment frequency, or expected environment of current missions better understand the linkages between climate and readiness and identify areas of risk that may require action in order for DoD to preemptively adapt to future impacts. Areas where the framework shows few links between hazards and readiness inputs may indicate where further investigation of causal mechanisms between climate and readiness is needed. Climate hazards that are implicated in many pathways may be of greater concern to future readiness than hazards that have only a few associated pathways, pointing to a need for early mitigation of climate risk and forward-looking, systemwide adaptation plans. Similarly, impacts on readiness inputs that are associated with a large number of pathways may reveal parts of the readiness-generation enterprise that are especially vulnerable to additive effects across multiple smaller disruptors. In the next chapter, we explore some of these patterns in detail.

CHAPTER 4

Climate Hazard Pathways: Populating the Climate Readiness Framework

The climate hazard framework introduced in the last chapter was built upon a set of *climate hazard pathways*, which describe the causal relationships between given climate hazards and their effects on specific readiness inputs. These pathways reflect global readiness effects under the current strategy and threat environment, assuming conditions that could occur under longer-term climate scenarios (present day to 2050). The climate hazard pathways were populated with information gathered during our discussions with subject-matter experts and an extensive literature review, as described in Chapter 2.

Visualizing Climate Hazard Pathways

As mentioned, a climate hazard pathway is a plausible avenue through which climate change could lead to a change in readiness outcomes in the absence of mitigation or adaptation. A climate hazard pathway

- Identifies the relevant climate hazard and any second-order effects that may result from the hazard. The hazards occur over land and water and at differing scales, frequencies, and intensities across the globe. They include those hazards that may occur gradually over time or may be distinct, extreme weather events.
- Provides a brief description of how the climate hazard interacts with the readiness input and enumerates the relevant impact (or impacts) on the readiness input caused by exposure to the climate hazard (e.g., reduced number of days during which daytime training is feasible), categorized by impact type (e.g., training [time] effects).
- Identifies the relevant readiness input (e.g., training) and input subcategories (e.g., field and outdoor) implicated in the climate hazard pathway.
- Lists possible adaptation measures that could mitigate the impact on readiness (e.g., use of simulated training environments).
- Includes historical examples of the hazard and its effects, where available.

Our original conceptualization of pathways was intended to include thresholds at which a particular pathway might become a concern for joint force readiness. However, our set of pathways does not include such thresholds, for two reasons. First, identification of applicable metrics for each pathway, as well as judgment about when impacts would become "significant," represented a significant amount of work for each pathway. We chose to focus on collection of pathways and iterative framework development instead of building thresholds for a smaller set of pathways. Second, we discovered that thresholds might not be pathway-specific, with the same readiness impacts potentially stemming from multiple hazards, meaning that thresholds and cumulative impacts must really be considered at a higher level than the individual pathway.

Figure 4.1 presents an example of five heat-related narratives that encompass both individual extreme heat events and gradual heat-related changes, such as rising average temperatures. These five narratives actually comprise six distinct pathways, since a pathway is characterized as having a single hazard and a single readiness input; *changing ocean conditions cause turbine failure* thus represents two pathways (equipment and force projection). Examining these narratives, we see, for example, that higher average temperatures (a hazard) can increase the geographic range and seasons habitable for mosquitoes, ticks, fleas, and sandflies, which increases the spread of vector-borne illnesses, such as malaria, dengue fever, and Zika, resulting in deleterious health effects (a readiness impact) and potential reductions in warfighter availability (a readiness input). Similarly, increasing ocean temperatures can stress equipment, such as turbines, that can overheat if not managed. The three additional narratives shown in the figure describe distinct, extreme heat events that can increase demand for energy, straining the electrical grid and leading to a suspension of activities; cause ambient temperatures to exceed safe thresholds, delaying training ("black flag" days); or damage infrastructure, leading to impacts on logistics activities.

We identified 110 pathways by which climate change could affect one or more readiness inputs at some point in the future.[1] In general, these pathways are broad descriptions of the ways in which climate hazards may influence readiness inputs. This set of pathways is by no means exhaustive; it represents both documented cases of climate-driven hazards affecting one or more inputs to readiness and plausible mechanisms by which such impacts could occur for which we were able to find supporting documentation. Thus, it represents a set of possibilities rather than a dataset of all mechanisms. Despite these limitations, this set of pathways was sufficient to support the development of the climate readiness framework and helps illuminate the relationship between climate and readiness in a variety of contexts.

While most pathways are not specific to a particular region or location, a few are driven by hazards found primarily in certain environments, such as sea ice melt and permafrost thaw. Likewise, some pathways are

FIGURE 4.1
Example Pathways for Heat

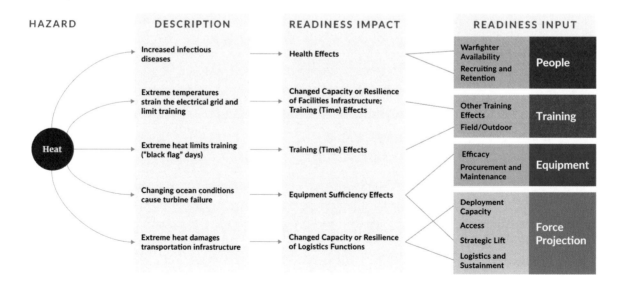

Sample Climate Hazard Pathways Illustrate the Revised Framework

---

[1] In total, we identified 63 distinct ways, or narratives, in which hazards could affect readiness. Many of these narratives include more than one hazard. Therefore, we identified a total of 110 unique pathways from a single hazard to a single readiness input.

very general in terms of the aspect of readiness affected (e.g., training), while others are more specific (e.g., cargo aircraft, frigates). Future analysis should more explicitly consider the locations at which impacts could occur, since the exposures of readiness inputs to these hazards, including their frequency, severity, and future projections, are specific to location. The hazards include gradual and lasting changes; such exposures will be enduring, and the impacts may be felt over time (*stressors*). Others are distinct events that occur at a particular frequency and severity (*shocks*); effects of such exposures may range from short-lived to permanent depending on their frequency and severity. Therefore, the pathways are not equivalent in terms of their potential effects on readiness.

The pathways that we identified are almost exclusively about threats to readiness. Both in the literature and in our discussions, there was minimal consideration of potential opportunities afforded by climate change. This bias could be a result of the characterization of climate effects as *hazards*, or the practice of considering threats to MA to anticipate and effectively mitigate the associated risks. One notable example of an opportunity is that sea ice retreat in the Arctic is opening new, shorter maritime routes, enabling greater maritime access to formerly ice-covered areas while also creating new requirements for force capabilities and capacity, greater domain awareness, and maritime surveillance in the region. However, because this was the only example of an opportunity identified, we limited our pathways to those that characterize negative consequences. The full dataset of narratives that describe the ways climate hazards can affect readiness is presented in a separate online appendix to this report.[2]

Our framework and the identified pathways describe the ways in which a climate hazard affects a readiness input and involve complex, dynamic processes compounded by the uncertainties around both future climate hazards and the readiness-generation process. Visualizations can be useful to analysts and decisionmakers for increasing awareness and knowledge of these issues, developing a greater understanding of these processes and the sources of uncertainties, and improving the likelihood that quantitative models will be accepted and utilized to catalyze action.[3] We developed several visualizations to enable analysts to probe different aspects of the pathways dataset and aggregate and visualize pathways within a standardized framework:

1. *static network diagrams*, focusing on sequential relationships between hazards, impacts, and readiness inputs
2. an *interactive network diagram*, enabling dynamic filtering of the network and examination of the details of individual pathways
3. an *interactive Sankey diagram*, depicting the volume of climate hazard effects as flows through a network[4]
4. *heat maps*, illustrating the frequency with which specific hazards, impacts, and readiness inputs appear in the pathways and highlighting patterns of occurrence of particular combinations of these elements.

Figure 4.2 illustrates as a network the subset of 22 pathways driven solely or in part by heat, where the nodes of the network are the stages of the pathways (i.e., hazards, second-order effects, readiness impacts, and readiness inputs) and the links are the connections between them. Each pathway is depicted using exactly one line in order to enable visual tracking of a single pathway from hazard to readiness input and to illustrate

---

[2] Best et al., 2023.

[3] Dominik Sacha, Hansi Senaratne, Bum Chol Kwon, Geoffrey Ellis, and Daniel A. Keim, "The Role of Uncertainty, Awareness, and Trust in Visual Analytics," *IEEE Transactions on Visualization and Computer Graphics*, Vol. 22, No. 1, 2016.

[4] To obtain a copy of the interactive network and Sankey diagrams as an HTML document, contact the authors.

FIGURE 4.2
Network Diagram of Heat Pathways

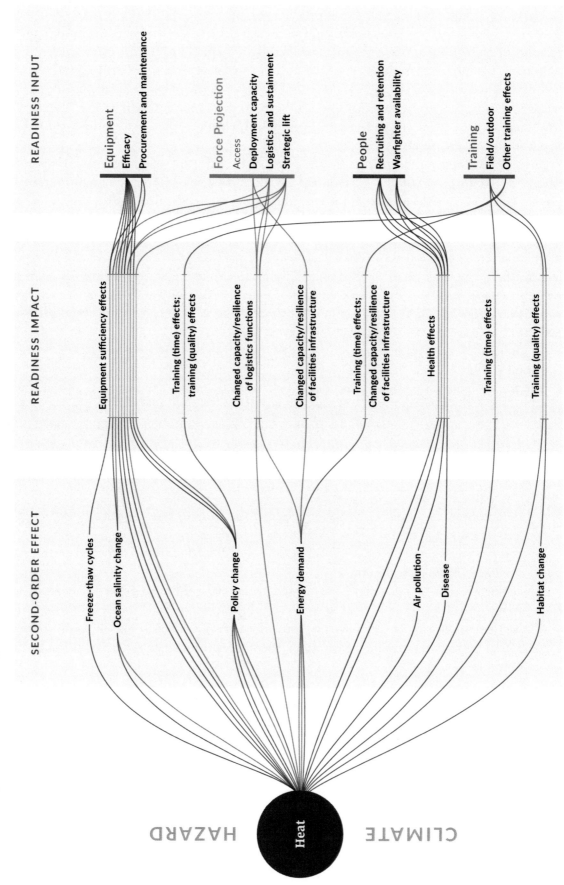

the density of the count of pathways passing through each node. Pathways split into multiple lines as they approach the readiness-input node if they affect more than one readiness-input subcategory. Color coding enables clear association of each pathway with its respective readiness input. In addition, the variation in the color of the pathways passing through each node provides a visual indication of the heterogeneity of readiness inputs associated with each node. For example, for heat-driven pathways, the *policy change* second-order effect receives pathways associated with *equipment* (blue), *force projection* (green), and *training* (red), while the *air pollution* and *disease* second-order effects are associated with *people* (purple) only. Pathway network diagrams for storms, flooding, and wildfire are shown in Appendix A.

In addition to the static diagrams, we developed an interactive version of the pathways network diagram that facilitates examination of the specific mechanisms of individual pathways and provides several dynamic filtering options. Users of the diagram may select filtering options from a drop-down menu to view subsets of the pathways by hazard, impact, or readiness input or may highlight individual pathways. Figure 4.3 depicts three views of this interactive diagram: (A) all pathways, (B) heat pathways only, and (C) a single pathway describing the impact of heat on time available for field and outdoor training (i.e., black flag conditions). Individual pathways highlighted in this way can be moved independently from the rest of the network by mouse-clicking and dragging in order to depict the connections between these nodes more clearly. Users may also look up a textual description of individual pathways by searching the pathways dataset for a particular pathway number or for keywords associated with the pathway (e.g., *heat* or *training*). An important limitation of this diagram is that links between nodes do not convey information about the number of pathways represented by the link. Rather, the diagram is intended to facilitate preliminary exploration of the types of pathways that involve certain hazards, impacts, and inputs and provide a platform to quickly obtain descriptive information about individual pathways of interest.

A third method of visualizing the pathways as a network enables more-direct comparison between nodes in terms of the volume of pathways passing through them. Figure 4.4 is a static image of an interactive Sankey diagram we developed that illustrates the 110 pathways we identified. In this type of diagram, the links, or *flow lines*, represent groups of connections between the five columns in the diagram, which represent hazards, second-order effects, readiness impacts, readiness-input subcategories, and readiness-input nodes. Both the number and the width of the flow lines convey information. Specifically, the number of flow lines indicates unique groupings of connections between the stages of the framework, while the line thickness represents the number of individual connections within each grouping as described by our pathways, where thicker lines indicate more connections. The diagram in its interactive form allows analysts to focus on hazards of concern, readiness impacts of concern, or readiness inputs to understand the extent of climate threats that the readiness system may be exposed to and the potential consequences to readiness.

As illustrated in the Sankey diagram, some hazards propagate effects through more-complicated flows, or *webs* (i.e., they may have second-order effects that could potentially affect a readiness input in several ways or affect multiple readiness inputs), suggesting the types of potential challenges to determining and managing these risks. For example, health effects, which has nine incoming flows from the climate hazards (primary hazards and second-order effects) and two outgoing flows to readiness-input subcategories, perhaps unsurprisingly is always associated with the *people* readiness input. Although determining the type and severity of exposure affecting the health and well-being of the force is somewhat complicated because of the nine different hazards that can drive such pathways, monitoring and responding to these effects should be comparatively straightforward because they are concentrated on a single readiness input (health effects affect readiness through warfighter availability and recruiting and retention). In contrast, facilities infrastructure is affected by 11 incoming hazard flows that are thicker than the health effects flows (average connections per flow line are 4.5 and 2.4, respectively) and represent every hazard except sea ice retreat, ocean anomalies, and other precipitation. In addition, pathways that involve damage to facilities infrastructure ultimately

FIGURE 4.3
Images of Interactive Network Depicting Pathways for All Hazards, Heat, and Black Flag Days

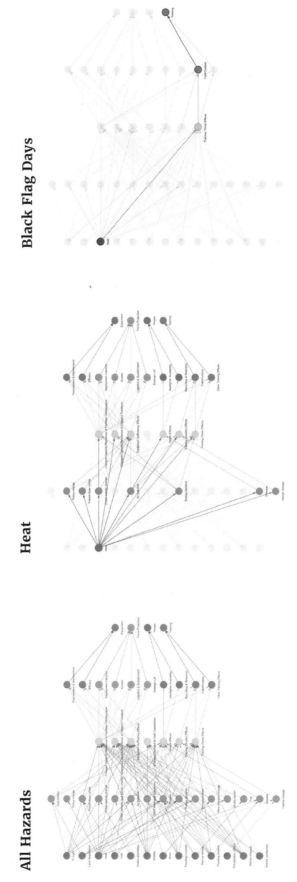

FIGURE 4.4
Image of Interactive Sankey Diagram of All Climate Hazard Pathways

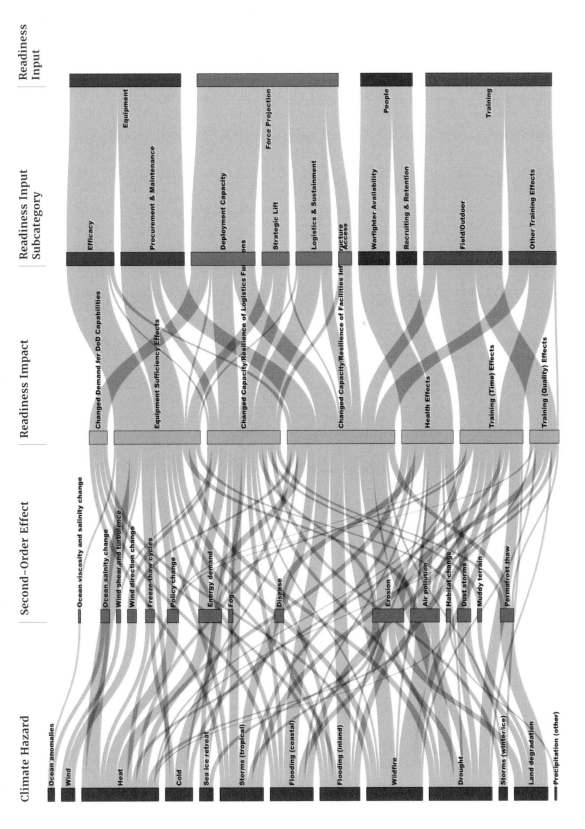

implicate three of the four readiness inputs (equipment, training, and force projection) along eight outgoing flows. This reinforces previous observations that facilities infrastructure should be a focal point for further study of readiness impacts of climate change and that differentiating exposure by service and facility function is essential for assessing the risk to individual readiness inputs.

A final visualization method compiles information from the pathways dataset into colorized tables, or *heat maps*, that highlight relative differences in the frequencies with which hazards, impacts, and readiness inputs appear in the pathways. Tables 4.1–4.3 show the frequency of occurrence in the pathways dataset of all pairwise combinations of (1) hazards and impacts, (2) hazards and readiness inputs, and (3) impacts and readiness inputs, respectively, where darker shading indicates higher frequency. The total number of occurrences of all combinations in each table is not equal to the total number of pathways, because many pathways have multiple impacts, each of which is counted separately in the tables. With this visualization method, it is immediately apparent which combinations of hazards, impacts, and/or readiness inputs have the greatest representation in the pathways dataset. For example, while heat was found to be a driver of nearly every type of impact identified, it is most frequently associated with equipment sufficiency effects (Table 4.1). Other hazards have impacts that are concentrated in specific areas of the readiness system; one example is sea ice retreat, which affects only the *force projection* readiness input (Table 4.2).

The pattern detection enabled by the visualizations is useful for identifying the most common (and uncommon) relationships between climate and readiness, as well as the distribution of the possible impacts caused by (or readiness inputs associated with) a given hazard. The visualizations may be useful for quantifying the frequency of elements within pathways as a first step toward prioritizing actions to address impacts on readiness. For example, one may first examine the relative importance of the hazards themselves: Are hazards that appear frequently in the pathways necessarily the most concerning for readiness? Because climate hazards differ in their geographic prevalence, severity (i.e., being nuisances versus having catastrophic impacts), and duration (i.e., gradual stressors versus punctuated shocks), focusing on simple counts of hazards in pathways alone can mask the true impacts of those hazards and may downgrade the importance of hazards that appear relatively infrequently but have significant effects. Likewise, a readiness input appearing in many pathways indicates numerous "opportunities" for its disruption, but not the full extent of its disruption by any one hazard. Thus, a high frequency of occurrence of a given combination of a hazard and an impact (or a readiness input) in the pathways dataset is not a direct proxy for the level of importance one should assign to that combination of hazard and impact (or readiness input). Additional qualitative analysis is required to interpret the information captured in these tables and other visualizations for prioritizing action to address climate risk to readiness.

Climate Hazard Pathway Prioritization Assessment

Through the climate readiness framework described in this report, we identify and assess relationships between hazards, impacts, and readiness inputs. However, the framework does not illustrate which readiness inputs might be more susceptible to impacts from climate hazards and which ones might require prioritization for action or further analysis. Because the climate hazard pathways do not include the ultimate outcomes or consequences to readiness, a definitive rank ordering is not possible using our framework. However, we can inform readiness planning by considering the individual determinants of overall risk. A risk calculation often considers two factors: the level of hazard exposure and the vulnerability to that hazard exposure. Vulnerability is often expressed as a combination of the sensitivity of the readiness input to the hazard and the adaptive capacity that exists to moderate the consequence of the hazard. By examining the individual determinants, we can reveal those that have high exposure or vulnerability; identify the key assumptions that drive the assessment, particularly where there is disagreement; and establish broad cate-

TABLE 4.1

Heat Map of Hazards and Impacts

Hazard	Changed Capacity and Resilience of Facilities Infrastructure	Changed Capacity and Resilience of Logistics Functions	Changed Demand for DoD Capabilities	Equipment Sufficiency Effects	Health Effects	Training (quality) Effects	Training (time) Effects	Total
Cold	3	2	0	4	0	0	3	12
Drought	6	4	0	4	5	3	6	28
Coastal flooding	7	2	2	2	1	1	4	19
Inland flooding	6	2	2	2	3	0	3	18
Heat	4	4	0	13	7	3	3	34
Land degradation	6	4	0	1	2	2	0	15
Ocean anomalies	0	0	0	3	0	0	0	3
Precipitation (other)	0	0	0	0	0	1	0	1
Sea ice retreat	2	4	0	0	0	0	0	6
Tropical storms	4	4	2	2	2	1	4	19
Winter/ice storms	1	2	0	0	0	0	1	4
Wildfire	7	1	2	6	3	2	4	25
Wind	2	3	0	1	0	0	0	6
Total	48	32	8	38	23	13	28	190

TABLE 4.2
Heat Map of Hazards and Readiness Inputs

Hazard	Equipment	Force Projection	People	Training	Total
Cold	4	3	0	5	12
Drought	7	5	5	11	28
Coastal flooding	5	6	1	7	19
Inland flooding	5	5	3	5	18
Heat	9	11	7	7	34
Land degradation	1	9	2	3	15
Ocean anomalies	2	1	0	0	3
Precipitation (other)	0	0	0	1	1
Sea ice retreat	0	6	0	0	6
Tropical storms	5	5	2	7	19
Winter/ice storms	0	2	0	2	4
Wildfire	9	5	3	8	25
Wind	2	4	0	0	6
Total	49	62	23	56	190

TABLE 4.3

Heat Map of Impacts and Readiness Inputs

Impact	Equipment	Force Projection	People	Training	Total
Changed capacity and resilience of facilities infrastructure	13	20	0	15	48
Changed capacity and resilience of logistics functions	5	27	0	0	32
Changed demand for DoD capabilities	0	8	0	0	8
Equipment sufficiency effects	31	7	0	0	38
Health effects	0	0	23	0	23
Training (quality) effects	0	0	0	13	13
Training (time) effects	0	0	0	28	28
Total	49	62	23	56	190

gories of pathways to help prioritize or guide readiness-planning activities that seek to reduce the effects of these hazards on readiness in the future.

To assess prioritization of climate pathways and readiness inputs, we conducted a modified Delphi exercise in which seven subject-matter experts from the RAND project team rated readiness-input and readiness-impact pairings (e.g., "people: health effects") based on their adaptive capacity and sensitivity to climate impacts.[5] The 110 pathways were grouped by readiness input and readiness impact to limit the demands on participants, but this process could be repeated at greater levels of detail if desired.

The process for our modified Delphi exercise was as follows: (1) Participants completed a survey to assign a rating for each readiness-input and readiness-impact pairing for adaptive capacity (*high, medium, low*) and sensitivity (*minimal, moderate, major*), including optional comments describing the basis for the ratings; (2) participants discussed the results of the ratings, specifically focusing on readiness inputs with divergence in the responses;[6] and (3) participants revised their ratings as needed based on the discussion. In a traditional Delphi exercise, the process is repeated until consensus is reached. For our purposes, one iteration (i.e., inputs, discussion, and revisions) was sufficient.

The survey instrument identified the specific hazards and pathways that contributed to each readiness-input and -impact pairing. Participants used our visualizations to understand exposure, which was helpful for contextualizing their vulnerability assessment. A rubric was provided to assist the participants and support more-consistent responses. The survey rubric and sheet are provided in Appendix B.

The rubric for adaptive capacity considers whether there are options for mitigating the climate hazard's effects on the readiness input. *High adaptive capacity* is when there are one or more low-cost, feasible options to reduce disruptions or impacts from the hazard. In contrast, *low adaptive capacity* is when there are few or no options to prevent disruption from occurring or when those that exist require substantial resources (i.e., financial, personnel, or time). Sensitivity ratings consider the extent to which the readiness input experienced effects after nominal exposure. *Low sensitivity* is when nominal exposure results in effects that are limited in scope or time, while *high sensitivity* is characterized as severe, widespread, or permanent damage or loss of effectiveness after nominal exposure.

During the assessments and discussions for adaptive capacity, we considered such factors as whether there are options for adjusting the readiness input, the resources (e.g., time or cost) involved to implement those options, whether the options rely on overall system capacity or redundancies, and whether the options present minimal or significant delay. For sensitivity, we considered how quickly the readiness input is affected by the climate hazard, to what extent nominal exposure is likely to affect function or efficacy during the period of exposure, and how quickly or how much effort it takes for the readiness input to recover. For both the adaptive capacity and sensitivity, ratings were agnostic to the location or geography where hazards could occur. Results from the ratings for adaptive capacity and sensitivity following the round of discussion are shown in Tables 4.4 and 4.5, respectively.

Our purpose was to understand how the pathways could be categorized at a high level to prioritize planning and investment activities. For a given exposure, pathways are considered vulnerable either because they are highly susceptible to the hazard (high sensitivity) or there are few options for adapting to the consequence of exposure (low adaptive capacity). We first identified readiness impacts for which survey partici-

---

[5] A Delphi exercise is a structured elicitation process used for forecasting or policymaking in a variety of areas, ranging from health care and education to defense strategic planning. It allows a panel of experts to reach a consensus on one or more discussion points through iterative rounds of elicitation and discussion.

[6] In this context, *divergence* is when at least one pair of opposite ratings exists among the other ratings. For example, if one analyst rated adaptive capacity as *high* and one analyst rated adaptive capacity as *low*, we would consider this a readiness input with divergence in the responses.

TABLE 4.4

Adaptive Capacity Ratings for Readiness Impacts

Readiness Impact	Readiness Input and Description	Rating								
Training (time) effects	(Training: Natural environment) Training time is reduced due to altered natural environment	Low to medium to high	Low	Low	Medium	Medium	Medium	Medium	High
Training (quality) effects	(Training: Facilities) Training quality is reduced due to damage to facilities	Low to medium to high	Low	Low	Medium	Medium	Medium	Medium	High
Training (time) effects	(Training: Facilities) Training time is reduced due to damage to facilities	Low to medium to high	Low	Low	Medium	Medium	Medium	High	High
Training (quality) effects	(Training: Natural environment) Training quality is reduced due to altered natural environment	Low to medium	Low	Low	Medium	Medium	Medium	Medium	Medium
Changed capacity and resilience of facilities infrastructure	(Equipment: Facilities) Equipment facilities are damaged	Low to medium	Low	Low	Medium	Medium	Medium	Medium	
Changed capacity and resilience of facilities infrastructure	(Force projection: Facilities) Deployment capacity is reduced due to damage or reduced access to facilities	Low to medium	Low	Low	Low	Medium	Medium	Medium	Medium
Equipment sufficiency effects	(Equipment: Sufficiency) Equipment itself is damaged or rendered less effective	Low to medium	Low	Low	Low	Medium	Medium	Medium	Medium
Equipment sufficiency effects	(Force projection: Equipment) Strategic lift capacity is reduced as a result of damaged or ineffective equipment	Medium	Medium	Medium	Medium	Medium	Medium	Medium	Medium
Changed capacity and resilience of logistics functions	(Force projection: Energy and water requirements) Readiness is reduced as a result of increased energy and/or water requirements of deployed forces	Medium to high	Unknown	Medium	Medium	Medium	Medium	Medium	High
Changed capacity and resilience of facilities infrastructure	(Force projection: Energy requirements) Readiness is reduced as a result of increased energy requirements of facilities	Medium to high	Medium	Medium	Medium	Medium	Medium	High	High
Changed capacity and resilience of logistics functions	(Force projection: Natural environment) Logistics capacity is reduced, disrupted, and/or more dangerous due to altered natural environment	Medium to high	Medium	Medium	Medium	Medium	High	High	High
Changed demand for DoD capabilities	(Force projection: Demand) Deployment capacity is reduced due to increased demand for HADR missions	Medium to high	Medium	Medium	Medium	High	High	High	High

Table 4.4—Continued

Readiness Impact	Readiness Input and Description	Rating							
Changed capacity and resilience of logistics functions	(Force projection: Infrastructure) Deployment capacity is reduced due to damage to supply chain and/or transportation infrastructure	Medium to high	Medium	Medium	Medium	High	High	High	High
Health effects	(People) Warfighter availability and/or ability to recruit is reduced	Medium to high	Medium	Medium	Medium	High	High	High	High
Changed capacity and resilience of logistics functions	(Equipment: Logistics) Equipment is damaged or rendered less effective due to damaged or ineffective supply chains and/or navigation routes	Medium to high	Medium	Medium	High	High	High	High	High

TABLE 4.5
Sensitivity Ratings for Readiness Impacts

Readiness Impact	Readiness Input and Description	Rating							
Changed capacity and resilience of logistics functions	(Equipment: Logistics) Equipment is damaged or rendered less effective due to damaged or ineffective supply chains and/or navigation routes	Minimal to moderate	Minimal	Minimal	Moderate	Moderate	Moderate	Moderate	Moderate
Changed capacity and resilience of logistics functions	(Force projection: Energy and water requirements) Readiness is reduced as a result of increased energy and/or water requirements of deployed forces	Minimal to moderate	Unknown	Minimal	Minimal	Minimal	Moderate	Moderate	Moderate
Changed demand for DoD capabilities	(Force projection: Demand) Deployment capacity is reduced due to increased demand for HADR missions	Minimal to moderate	Minimal	Minimal	Minimal	Minimal	Minimal	Minimal	Moderate
Changed capacity and resilience of logistics functions	(Force projection: Natural environment) Logistics capacity is reduced, disrupted, and/or more dangerous due to altered natural environment	Moderate	Moderate	Moderate	Moderate	Moderate	Moderate	Moderate	
Health effects	(People) Warfighter availability and/or ability to recruit is reduced	Moderate to major	Moderate	Moderate	Moderate	Moderate	Moderate	Moderate	Major
Training (time) effects	(Training: Facilities) Training time is reduced due to damage to facilities	Moderate to major	Moderate	Moderate	Moderate	Moderate	Moderate	Major	Major
Equipment sufficiency effects	(Force projection: Equipment) Strategic lift capacity is reduced as a result of damaged or ineffective equipment	Moderate to major	Moderate	Moderate	Moderate	Moderate	Moderate	Major	Major
Changed capacity and resilience of logistics functions	(Force projection: Infrastructure) Deployment capacity is reduced due to damage to supply chain and/or transportation infrastructure	Moderate to major	Unknown	Moderate	Moderate	Moderate	Moderate	Major	Major

Table 4.5—Continued

Readiness Impact	Readiness Input and Description	Rating							
Training (quality) effects	(Training: Natural environment) Training quality is reduced due to altered natural environment	Moderate to major	Moderate	Moderate	Moderate	Moderate	Major	Major	Major
Training (time) effects	(Training: Natural environment) Training time is reduced due to altered natural environment	Moderate to major	Moderate	Moderate	Moderate	Moderate	Major	Major	Major
Training (quality) effects	(Training: Facilities) Training quality is reduced due to damage to facilities	Moderate to major	Moderate	Moderate	Moderate	Moderate	Major	Major	Major
Equipment sufficiency effects	(Equipment: Sufficiency) Equipment itself is damaged or rendered less effective	Moderate to major	Moderate	Moderate	Moderate	Major	Major	Major	Major
Changed capacity and resilience of facilities infrastructure	(Equipment: Facilities) Equipment facilities are damaged	Moderate to major	Moderate	Moderate	Moderate	Major	Major	Major	Major
Changed capacity and resilience of facilities infrastructure	(Force projection: Facilities) Deployment capacity is reduced due to damage or reduced access to facilities	Moderate to major	Moderate	Moderate	Moderate	Major	Major	Major	Major
Changed capacity and resilience of facilities infrastructure	(Force projection: Energy requirements) Readiness is reduced as a result of increased energy requirements of facilities	Minimal to moderate to major	Minimal	Minimal	Moderate	Moderate	Moderate	Major	Major

pants quickly converged toward a particular rating, indicating a consistent understanding of the implications of a climate hazard on the readiness input. For instance, the adaptive capacity of *force projection: energy requirements* and *force projection: energy and water requirements* was rated *medium to high*. The difference between a rating of *medium* and a rating of *high* may be dependent on such factors as potential technical constraints or electrical grid constraints but is relatively insensitive to changes in assumptions. The second group of readiness inputs consists of those for which survey participants diverged in their ratings, indicating that the assessment may vary widely depending on assumptions regarding the specific scenario considered. For example, the sensitivity of *force projection: energy requirements* could be rated *low, medium,* or *high* depending on assumptions regarding such factors as energy grid capacity, opportunities to leverage the commercial sector, whether effects are immediate or temporary, and variation in impacts to different installations and services. Areas of divergence can reveal which assumptions are critical for determining the impacts and may indicate that a more targeted assessment may be needed to prioritize action.

Adaptive Capacity

As Table 4.4 suggests, the readiness inputs of *people, force projection* (relating to transportation infrastructure), and *equipment* (relating to logistics functions) appear to have high adaptive capacity. In the case of *people*, there are many measures, including physical, behavioral, and procedural (such as rescheduling), that can be used to reduce the harmful effects of a climate hazard. In addition, in discussions with the readiness community, we heard that people tend to be adaptable even during difficult conditions.[7] Two other readiness inputs, *force projection* (relating to transportation infrastructure) and *equipment* (relating to logistics) were also considered relatively adaptable because there are choices among transportation modes, alternative routes available for providing supplies and equipment, or opportunities to make temporary adjustments to the use of equipment when the duration of the hazard is limited. In contrast, such readiness inputs as *force projection* (relating to facilities) and *equipment* (relating to sufficiency) were assessed as having low or medium adaptive capacity, in large part because there are limited, expensive options to mitigate hazards to facilities, especially when the urgency of operations' timing is a major consideration. Adaptive capacity for *equipment* (relating to sufficiency) depends on the type of equipment, but, in general, it was assessed to be limited, and, if exposure to hazards presents a "new normal," then design specifications will have to be revised. Our final observation is that there is a natural divide between those readiness inputs that are relatively fixed and long lived, such as training, force projection, and training facilities and equipment (sufficiency), and those that have more adaptive capacity, such as logistics functions or people (as can be seen in Table 4.4).[8]

Sensitivity

In terms of sensitivity to climate hazards, the readiness inputs of *force projection* (relating to demand for DoD capabilities), *force projection* (relating to energy and water), and *equipment* (relating to logistics) are not particularly susceptible to the climate hazards identified in the pathways (as shown in Table 4.5). Participants noted that the increase in *force projection* (relating to demand for DoD capabilities) for HADR likely will not require a substantial increase in demand for resources and assets and may be shared with partners and allies, and, since participation is driven by policy, there are opportunities to ensure that readiness impacts are limited. *Force protection* (relating to energy and water) was also assessed to be less sensitive to climate hazards because the increase in demand for these resources is not anticipated to be appreciably different from that of today and because there is capacity to increase supply. One notable exception where this readiness input could be more sensitive is at forward operating bases that experience frequent maneuvers. At the other end of the spectrum are *force projection* (relating to facilities), *equipment* (relating to facilities), and *equipment* (relating to efficacy), which are particularly sensitive to the hazards identified in the pathways. In large part, facilities are considered sensitive because of their age, because of design standards that are derived from historical climate and weather patterns, and because military facilities are generally not considered to be well maintained.[9] And although equipment sensitivity may vary by equipment type and associated design thresholds, overall equipment was assessed as being sensitive to the climate hazards because it is designed in accordance with performance requirements that are fixed in time and might not account for future climate change. Furthermore, there are examples from historical experience of hazards degrading equipment and affecting operations. A summary of participant comments for each of the readiness-input and -impact pairings presented in Tables 4.4 and 4.5 is included in Appendix B.

The information generated by a full Delphi analysis that includes both exposure and vulnerability could be used to categorize the potential readiness impacts to determine next steps for analysis and risk mitigation. Climate hazard risk mitigation generally involves steps to identify key hazards and assets; assess exposure, vulnerability, and impacts; develop mitigation strategies; implement these strategies; and monitor. This

---

[7] Navy personnel, discussions with the authors, December 8, 2021.

[8] International security expert, discussion with the authors, April 28, 2022.

[9] U.S. Government Accountability Office, *Defense Infrastructure: DOD Should Better Manage Risks Posed by Deferred Facility Maintenance*, GAO-22-104481, January 2022.

process is iterative or near continuous.[10] It shares many similarities with the construct DoD has laid out for MA, which includes the steps of *identification* of important factors, *assessment* of risk, development of *risk-management* actions, and *monitoring and reporting*.[11] The MA process is discussed in more detail in the next chapter as a possible way to integrate climate thinking with current DoD processes. An illustrative example of how the readiness impacts might be categorized to guide action is shown in Table 4.6.

Moving Beyond Pathway Patterns

In this chapter, we discussed some of the patterns of relationships among pathways observed in the current application of the climate readiness framework. While analysis of such patterns is helpful, it provides only a first step toward definitively identifying, prioritizing, or ranking either climate hazards or the impacts on readiness they might have. Qualitative analysis of sensitivity and adaptive capacity characteristics, of the sort demonstrated by our modified Delphi exercise, can illuminate implications and underlying assumptions of pathways and help move further toward prioritizing key areas for action. The current application of the climate readiness framework is populated by a mixture of plausible and observed pathways collected through discussions and review of literature, meaning that it will require updates as climate hazards evolve over time and may include some pathways that never become reality. Furthermore, some pathways are specific to a particular location or class of weapon systems and equipment, while others describe more-general effects. Therefore, the framework and the pathways that populate it stop short of offering quantitative assessments of the likelihood or level of impact of each link between climate and readiness. Further data collection, metric development, and analysis will be required to quantify such factors as likelihood and level of

TABLE 4.6

Categorizing Readiness-Input Impacts to Prioritize Action

Exposure	Vulnerability (equal weighting of sensitivity and adaptability)	Priority and Suggested Actions
High	High	• These high-priority pathways are where multiple actions and investments are likely needed. • Assess specific risks and vulnerabilities across locations or assets, develop indicators, and monitor immediately. • Incorporate indicators and analysis into the MA analysis process. • Begin investing in mitigation actions and investments.
Moderate	High	• These pathways are where multiple actions and investments are likely needed, but exposure is moderate. • Suggested actions are the same as for high-exposure, high-vulnerability pathways but focus on mission-essential, or critical, assets or locations.
High to moderate	Low to moderate	• Ensure that decisionmakers are aware of the vulnerabilities of affected readiness inputs and continue to mitigate as feasible.
Low	Moderate to high	• Monitor exposure.
Low	Low	• Periodically revisit exposure.

---

[10] Leidos, Inc., *Climate Change Planning Handbook: Installation Adaptation and Resilience*, prepared for Naval Facilities Engineering Command, January 2017; Pinson et al., 2020; U.S. Climate Resilience Toolkit, "Steps to Resilience Overview," webpage, last updated July 28, 2022; U.S. Department of Transportation, Federal Highway Administration, "Vulnerability Assessment and Adaptation Framework, 3rd Edition," webpage, last updated January 22, 2018.

[11] U.S. Department of Defense, *Mission Assurance Construct*, DoD Instruction 3020.45, May 2, 2022a.

exposure, severity of disruption, and available adaptive capacity. The patterns identified through the climate readiness framework shed light on the kinds of data and analysis that may be required to quantify risk for different kinds of climate readiness pathways, as we will discuss further in Chapter 5.

Applying the Climate Readiness Framework

The climate readiness framework and pathways that populate it, presented in the previous two chapters, provide a starting point for operationalizing links between climate risk and readiness and integrating those links with the existing readiness enterprise. In this chapter, we present a set of potential *climate readiness integration points* to be considered for further study or development by DoD and the services. Climate readiness integration points are potential ways of incorporating climate risk into existing readiness processes, models, tools, assessments, and/or data structures. The integration points presented in this chapter were compiled through a combination of coincidental identification during our literature review and climate- and readiness-related outreach and a search for specific types of integration points that could help DoD capitalize on the work laid out in this report. This chapter describes a diverse set of integration points identified throughout the course of the study. These examples are not intended to be comprehensive, but they do provide a sampling of the kinds of ways that the climate readiness framework could be integrated with the existing readiness enterprise. The DoD Climate Adaptation Plan stresses the importance of actually "*Implementing* Climate-Informed Decision-Making" and highlights three main focus areas for achieving such implementation:

- incorporation of "Climate Intelligence" stresses the need to consider climate as a source of risk in all types of threat assessment
- incorporation of climate into "Strategic, Operational, and Tactical Decision-Making" highlights the need to consider climate when developing strategies and policies ranging from high-level warfighting concepts to specific budgeting and planning factors
- incorporation of climate into "Business Enterprise Decision-Making" focuses on the need to consider climate in enterprise-wide resourcing decisions and cost-benefit analysis.[1]

In the next sections, we detail the set of identified potential integration points that speak to all three of these focus areas, illustrating how their further development could help spur the kind of "implementation" suggested in the Climate Adaptation Plan.

Building Climate Intelligence into Mission Assurance Processes

Given the complexity of the set of challenges that DoD and the joint force face in the global security environment, DoD requires a comprehensive and integrated process to assess and address risk in the performance of mission-essential functions, to deconflict and reduce programmatic redundancy, and to identify effective risk-mitigation programs or resource allocations in a fiscally constrained environment. In 2012, DoD designed such a comprehensive process—MA—with the intent of institutionalizing

---

[1] U.S. Department of Defense, 2021b, p. 6, emphasis added.

[a] process to protect or ensure the continued function and resilience of capabilities and assets—including personnel, equipment, facilities, networks, information and information systems, infrastructure, and supply chains—critical to the performance of DoD [mission-essential functions] in any operating environment or condition.[2]

Accordingly, MA is intended to provide an integrative and common framework within which to broadly consider the impacts of all complex threats and hazards—both man-made threats and naturally occurring climate hazards. The MA process is envisioned to leverage existing protection and resilience programs and provide inputs to the overall DoD and military department planning, budgeting, requirement, and acquisition processes.

Consistent with the 2012 Mission Assurance Strategy, the Office of the Under Secretary of Defense for Policy published DoD Directive 3020.40, *Mission Assurance (MA)*, which "[e]stablishes policy and assigns responsibilities to meet the goals of refining, integrating, and synchronizing aspects of DoD security, protection, and risk-management programs that directly relate to mission execution as described in the DoD Mission Assurance Strategy and Mission Assurance Implementation Framework."[3] Similarly, this directive established a coordination-board structure to manage risk across all DoD components; ensure the incorporation of MA into Defense Planning Guidance; and integrate high priorities into the respective planning, programming, budgeting, and execution processes. The DoD components followed suit by publishing their respective supporting MA instructions and thus established a common framework for MA decisionmaking both within and across military departments.[4]

We reviewed the collective MA frameworks documented in the respective service instructions for guidance regarding general climate effects on overall mission accomplishment. We found the documents to be comprehensive and holistically oriented toward *all* threats and hazards, vice uniquely concentrated on or calling out climate effects. The guidance across services directed the execution of unified efforts for making risk-informed decisions, with specific emphasis on integrating, coordinating, synchronizing, and prioritizing various service programs, initiatives, and resources. The MA efforts tended to focus on risk management from a short-term perspective but also mandated a forward-looking perspective. The MA processes were designed to align with senior leaders' strategic goals while also maintaining flexibility to identify and meet emerging threats and hazards. Thus, the instructions concentrated on enhancing the services' comprehensive ability to prepare for, prevent against, respond to, and recover from any such negative incidents. As a result, MA presents itself as a promising vehicle for incorporating climate intelligence into existing threat-assessment activities.

As part of the MA assessment program, detailed in DoD Instruction 3020.45,[5] each service conducts a location-specific baseline AHTA. These assessments are conducted at the service installation level and serve as the assessment for all MA programs and activities to determine specific hazards and threats, ranging from natural events to human-caused events (accidental and intentional) and technologically caused events.

---

[2] U.S. Department of Defense, 2012, p. 1.

[3] U.S. Department of Defense, *Mission Assurance (MA)*, DoD Directive 3020.40, September 11, 2018b, p. 1.

[4] Department of the Air Force, *Mission Assurance*, Air Force Policy Directive 10-24, November 5, 2019; Department of the Army, *The Army Protection Program*, Army Regulation 525-2, December 8, 2014; Department of the Navy, *Navy Mission Assurance Program*, OPNAV Instruction 3502.8, November 8, 2017; Department of the Navy, Headquarters U.S. Marine Corps, *Marine Corps Mission Assurance*, Marine Corps Order 3058.1, October 23, 2014; Secretary of the Army, "Army Directive 2020-08 (U.S. Army Installation Policy to Address Threats Caused by Changing Climate and Extreme Weather)," memorandum for U.S. Army North, U.S. Army South, U.S. Army Africa/Southern European Task Force, et al., September 11, 2020.

[5] U.S. Department of Defense, 2022a.

The RAND project team reviewed the current 2021 AHTA methodology used to support the Marine Corps Base Camp Lejeune Mission Assurance Assessment in accordance with Marine Corps Order 3058.1.[6] The Marine Corps MA assessment team based its review on interviews with personnel and officials of the resident sections and organizations at the installation, direct field observations of the assessed concepts, and reviews of applicable documents. Using a compilation of sources, the assessment team characterized a variety of threats and hazards for Camp Lejeune, made objective ratings of location-specific vulnerability and relative risk for each threat and hazard based on explicitly stated criteria, detailed actionable recommendations, and then projected revised ratings for location vulnerability and risk based on the recommendations. The intended use of these results was to inform local commanders and their higher headquarters (including various other Marine Corps decisionmaking forums) about potential risk-reduction measures or investments and relative contributions to risk and vulnerability reduction.

The comprehensive methods employed in this Marine Corps MA assessment for Camp Lejeune were explicit in terms of their approach and data sources, reproducible by other professionals, and published for external review. This degree of structure, discipline, and transparency provides a sound basis for considering climate effects and hazards affecting Marine Corps mission-essential functions.

Although certainly open to further review and debate, the applied AHTA approach did allow all hazards and threats to be considered on the same scale and thereby assessed equitably in establishing the relative prioritization of recommended courses of action. As discussed in the previous chapter, the steps involved in the overall MA construct have some parallels with the way experts think about climate hazard adaptation. In this way, climate impacts on MA can be directly compared with other threat considerations that degrade MA—a quality that senior leaders frequently lack in their decisionmaking deliberations for diverse recommendations that are strongly supported by divergent stakeholders. The MA AHTA approach, if implemented over a sufficient period, would allow for the systematic consideration of trends affecting mission-essential functions across the services and local installations, including long-term climate change, thereby allowing DoD to integrate climate intelligence into strategic risk issues through an existing process. Finally, the analytical process considered a variety of alternative solutions, and not all recommendations necessarily involved additional fiscal resources. Recommendations that require limited or no additional fiscal resources generally involve nonmateriel solutions, such as changes in operational procedures or concepts; an alternative use of existing assets or human resources; or policy modifications.[7]

Integrating Climate Risk into Existing Decision-Support and Planning Tools

Across DoD and the services, many large and small processes continuously analyze data and qualitative inputs to make decisions at the tactical, operational, and strategic levels. In the course of our study, we identified a small number of modeling tools that could incorporate inputs related to climate exposure in order to produce climate-informed outputs to inform existing readiness-related decisionmaking. We did not attempt to conduct an exhaustive review of existing DoD models of readiness-related processes that may be affected by climate, but we came across two cases that provide interesting examples of analytic processes that currently or could account for weather and climate effects. In this section, we describe these examples without offering specific recommendations for implementing the integration of climate-related effects. Rather, we

---

[6] Department of the Navy, Headquarters U.S. Marine Corps, 2014.

[7] The analyses supporting such determinations are developed from doctrine, organization, training, leadership and education, personnel, facilities, and policy (DOTMLPF-P) analyses associated with capabilities-based assessments.

offer these examples to highlight that existing analytic processes could be used—in some cases, in modified forms—to inform strategic-readiness assessments on the potential effects of future climate change.

Seaport Capacity

The first example involves the periodic series of *Port Look* studies conducted by SDDC's Transportation Engineering Agency. In addition to the *Port Look* studies focused on seaports needed to move DoD unit cargo, separate studies focus on ammunition ports.

The *Port Look* studies examine the current capacity of strategic seaports and alternate seaports to meet modeled throughput for major deployments, such as to execute combatant command operational plans.[8] The methodology of the studies is to

> assess current seaport capabilities and coastal throughput requirements, *account for threats that could have an impact on military seaport deployment operations*, then assess if the current capacities within the Strategic and Alternate Seaport portfolios is sufficient to ensure the timely delivery of Department of Defense (DOD) unit equipment cargo.[9]

In terms of possible threats to port capacity, the major climate hazard considered by current modeling is hurricanes. The studies use historical data to determine the possibility that hurricanes or other major disaster events (e.g., earthquakes, volcanic eruptions, tsunamis) could limit DoD's ability to use one or more seaports at the time of a major deployment. While the studies do incorporate up-to-date data on the frequency of hurricanes, they do not attempt to project future frequency or severity of hurricanes due to climate change. They also do not explicitly consider other potential effects of climate change, such as sea-level rise, though they are able to include updated data on port access. Most port facilities that DoD would use in a major deployment are commercially owned. The basic assumptions that the Transportation Engineering Agency employs are that (1) the owners of such ports will be motivated to keep these ports commercially viable and will make needed investments to mitigate the dangers associated with sea-level rise and (2) if the owner of a given port does not, there is sufficient capacity in the U.S. port system overall for DoD to shift operations to a different port.[10] This latter assumption is backed by the premise that, because there are no specific investments that DoD is going to make in commercial ports anyway, monitoring commercial port capacity overall is sufficient. SDDC monitors all of the ports in the strategic and alternate seaport portfolios and has found that it would have time to shift focus to a different port if one became unviable as a result of sea-level rise or another factor.

Of greater concern is ammunition movement. There are two military ocean terminals (one on each coast) and a West Coast naval magazine that account for most shipments of ammunition. These are not included in the *Port Look* studies, but they are considered in separate analyses, and there are processes to account for sea-level rise there, as part of sustainment investment decisions.[11]

---

[8] These studies are conducted "either (1) on an as needed basis when mandated by the U.S. Congress, or (2) after each major Mobility Capability and Requirements Study" (SDDC Transportation Engineering Agency, *Port Look 2021*, May 10, 2021, p. 11, Not available to the general public).

[9] SDDC Transportation Engineering Agency, 2021, p. 11, emphasis added.

[10] Government official from SDDC Transportation Engineering Agency, discussion with the authors, November 11, 2021. There are 18 primary ports in the strategic seaport program and 14 alternates—more than enough to meet the needs of a single major deployment.

[11] Government official from SDDC Transportation Engineering Agency, discussion with the authors, November 11, 2021; SDDC Transportation Engineering Agency, 2021, p. 20.

In addition to the above, SDDC has an annual data call with the geographic combatant commands and service components to identify any port-related issues that should be assessed in their respective areas of operation. That process is not by design intended to look at factors related to climate change, although an SDDC official expressed the belief that the process would highlight related challenges that should be addressed.[12]

Current approaches to analyzing port throughput already demonstrate how analytic processes can account for a small number of specific climate hazards in ways that could be used to inform strategic-readiness assessments. Our work identifies several pathways that could disrupt port capacity and throughput in ways that could both shock and stress the systems being modeled through *Port Look* and related studies. Wildfires, storms, and flooding might close down entire facilities after extreme events, while sea-level rise and increased heat might require infrastructure investments or result in higher energy demands and costs. The *Port Look* set of studies provides an opportunity for more explicitly exploring these risks, with frequent model runs and the ability to incorporate new data providing an avenue for modeling potential future impacts of climate change. While we did not identify specific ways in which current port assessment processes could be updated to account for a broader set of potential future impacts of climate change, existing capability to examine the impacts of closed facilities, reduced throughput, and the need to use alternate routes lay the groundwork for incorporation of these additional risk factors.

Training Range Capacity

The second example of existing analytic capabilities that could be updated to include climate change risk factors is related to training range capacity. The Army routinely evaluates the throughput capacity of its training ranges to assess adequacy to support training requirements. One tool used is the Army Range Requirements Model (ARRM). ARRM supports calculations about training capacity, throughput, and shortages.[13] More specifically, ARRM compares range throughput requirements with throughput capacity. Requirements are based on the number of soldiers or units that must execute training over time and the nature of that training. The *throughput capacity* of a range measures the number of soldiers or units that it can accommodate over the same period.

However, black flag periods caused by extreme heat, as well as other climate-related risk factors, may limit the hours that ranges are available for use, constraining throughput capacity. Our pathways include 21 ways that time available for training could be affected by climate hazards, such as storms, flooding, wildfire, cold, heat, and drought. Beyond impacts on training time, our pathways include six additional narratives that might affect quality of training, for example because of land degradation or the need to alter training to protect threatened habitats. When compared with requirements, a climate-informed assessment could highlight gaps in capacity where the current approach would identify no such gaps. Forecasts of potential future impacts to range throughput capacity due to exposure to adverse climate-related effects could be used to inform strategic-readiness assessments related to training and infrastructure.

Integrating Climate Risk into Assessment of Tactical, Operational, and Strategic Readiness

One of the purposes of readiness-related models and analyses, like those described in the previous section, is to inform readiness measurement and assessment, ranging from tactical to operational to strategic readi-

---

[12] Government official from SDDC Transportation Engineering Agency, discussion with the authors, November 11, 2021.

[13] Department of the Army, *The Army Sustainable Range Program*, Army Regulation 350-19, August 30, 2005.

ness. Information about impacts of climate hazards on specific components of readiness, such as Army training range capacity from the above example, must feed into standardized readiness metrics and assessment frameworks before it can inform higher-level decision processes. While a review of service and DoD readiness reporting and assessment is outside the scope of this report, this section provides a brief overview of current reporting architectures and comments on how readiness measurement and assessment may be updated to better incorporate some aspects of climate risk. Incorporating climate into readiness measurement and assessment speaks most immediately to the Climate Adaptation Plan's "Strategic, Operational, and Tactical Decision-Making" focus area, but it also moves toward informing enterprise decisionmaking because it allows readiness-related resources to be weighed against other department priorities.

The Chairman's Readiness Assessment considers joint readiness at three conceptual levels, as discussed in the introduction to this report: tactical, operational, and strategic.[14] In terms of process, it focuses on two:

- *unit-readiness reporting* (i.e., tactical-level readiness), which is reported through the Defense Readiness Reporting System (DRRS)
- *strategic-readiness assessments* (which combine the operational and strategic conceptual levels) as part of the Joint Combat Capabilities Assessment (JCCA); the JCCA includes a number of subprocesses and produces the Joint Force Readiness Review, the Readiness Deficiency Assessment, input to the Semiannual (formerly *Quarterly*) Readiness Report to Congress, and plan assessments, among other outputs.[15]

Both processes are focused on short-term readiness and do not attempt to consider the potential readiness impacts of factors—like climate change—that may play out over the course of years or decades. However, there are examples of special readiness assessments—such as the JCCA and SRRC examination of the readiness impacts of the coronavirus disease 2019 pandemic—that could be replicated for climate hazards. Alternatively, climate impacts on readiness could regularly be incorporated into the SRRC section on infrastructure, assets, and logistics.

Unit-Readiness Reporting (Tactical Readiness)

Tactical-level unit-readiness reporting is done through DRRS, which "provides a means to manage and report the readiness of the DoD and its subordinate Components to execute the National Military Strategy."[16] OSD maintains the DRRS-Strategic (DRRS-S) system. Some services maintain their own DRRS systems, which feed DRRS-S; for example, the Army uses DRRS-Army (DRRS-A).[17] Other services, as well as combatant commanders and defense agencies, report directly into DRRS-S.

The primary metrics reported through DRRS focus on personnel readiness (P-level), equipment on hand (S-level), equipment condition (R-level), and training status (T-level), which are used to derive an overall unit-readiness level (C-level). Moreover, DRRS has the ability to collect and report data on how various types of resource constraints affected the unit's readiness levels in a given reporting period. (These fields provide additional information and do not factor directly into calculations of the various readiness levels.) In addition, DRRS collects commanders' comments in free-text fields, allowing for a fuller and more nuanced pre-

---

[14] Chairman of the Joint Chiefs of Staff, 2010, pp. 9–10.

[15] For more information on the JCCA and its products, see Chairman of the Joint Chiefs of Staff, 2010, pp. 13–20.

[16] U.S. Department of Defense, *Department of Defense Readiness Reporting System (DRRS)*, DoD Directive 7730.65, May 31, 2018a, p. 1.

[17] Such systems must meet or exceed the minimum data standards for DRRS-S (U.S. Department of Defense, 2018a, p. 7).

sentation of the factors affecting a unit's readiness. However, DRRS does not provide a built-in capability to "mine" commanders' comments for data, which a user could then aggregate into summary reports (e.g., spanning multiple units or periods).

As noted above, unit-level reporting in DRRS focuses on short-term unit readiness. It does not attempt to forecast future unit readiness.[18] More importantly for this discussion, it does not have any standardized fields to report on how weather or climate hazards might have affected a unit's achieved readiness level in a given period (e.g., if a unit was impeded in executing its training plan because of a significant number of black flag days). It is possible that commanders' comments address such factors at times, but it seems unlikely that most commanders do so in a systematic way.[19]

Although there is no current reporting requirement related to weather and climate hazards, it seems highly likely that units must already manage the impacts of such hazards on their readiness. Heat, storms, flooding, wildfires, disease, and the like are not new, even if their prevalence should increase in the future with a warming climate. Modifying current unit-level reporting requirements to account for climate hazards could provide DoD with a valuable data source for understanding how climate affects tactical-level readiness. An initial approach could be to simply add a dedicated comment field for commanders to capture such effects. If such comments proved to be a rich source of data, then reporting requirements could be further modified so that such data could be captured more systematically to more readily support data aggregation.

Importantly, it may be that adverse weather and climate-related phenomena do not often create measurably adverse impacts on units' reported readiness levels—and this could continue to be the case in the future. This could be because commanders believe that they must build and maintain readiness despite adverse environmental conditions and either persist despite these conditions or look for ways to mitigate adverse impacts (e.g., nighttime training or the use of simulators to avoid daytime black flag periods) so that they do not significantly inhibit the unit's training progression. However, adaptations employed to moderate harm by adverse weather and climate hazards often come with a cost, financial or otherwise. A significant benefit of modifying readiness-reporting requirements in DRRS to account for climate hazards could be to help DoD monitor, assess, and forecast the costs of such adaptation measures on a more systematic basis. This ability may be particularly valuable because adaptation costs seem likely to increase with a warming climate, and their increasing magnitude may threaten to crowd out other activities that DoD must fund with a limited budget.

Installation-Readiness Reporting (Tactical Readiness)

In addition to unit-level reporting, DRRS captures readiness information supplied by installations. The mechanisms and exact contents of reporting differ by service. For example, Navy Installations Command is responsible for DRRS reporting on shore installations. Navy mission-essential task list information is required to be reviewed annually to ensure that it is accurate. DRRS-Navy (DRRS-N) is the authoritative system for Navy readiness data for personnel, equipment, supply, training, ordnance, and facilities, and assessments must use the Navy's personnel, equipment, supply, training, ordnance, and facilities (PESTOF)

---

[18] In some cases, such as in the U.S. Air Force, units do report projected readiness for three, six, and 12 months out, but these reports do not cover the relevant time frames for climate change impacts.

[19] Although we did not attempt to assess the extent to which climate information is captured in commanders' assessments, discussions with service stakeholders and subject-matter experts suggested that such information would likely be collected only sporadically.

data. The PESTOF system draws from authoritative data sources to supply assessment information on a unit's ability to perform tasks and provide capabilities needed to support operational forces.[20]

The Army source for DRRS-A reporting is the Installation Status Report (ISR) for infrastructure, which is a near-real-time reporting system that provides information on the condition of facilities, infrastructure, and natural resources, assessed against Army-wide standards. There are three ISR elements:

- *ISR-Infrastructure*, which reports on the condition ("capability, quality, and readiness") of infrastructure "compared to established Army standards" and, where applicable, includes costs to meet these standards
- *ISR-Mission Capacity*, which reports on the "current and future mission requirements, readiness, and . . . sustainment with critical resources such as air, land, water, and energy"
- *ISR-Services*, which reports on the "quality and quantity of installation services against Federal, Department of Defense (DOD), or Army standards."[21]

Only two components, ISR-Infrastructure and ISR-Services, support DRRS-A reporting. Installation mission-essential tasks are identified in ISR, and the information is used to generate ratings for mission support, quality, and readiness.[22] According to Army guidance, designated installations are required to submit readiness information quarterly and report status changes within 24 hours, such as after a major storm.[23]

According to DAF guidance, installation mission-essential tasks should include, at a minimum, "airfield operations; munitions supply, storage and distribution; petroleum, oil, and lubricants; provide contingency billeting; and range operations," and assessments should consider

- "New encroachment concerns or environmental impacts"
- "Natural disasters affecting installation operations for over 72 hours"
- "Legislative changes impacting training capabilities"
- "Infrastructure degradation or failing due to funding."[24]

Although not all climate hazards hit the readiness inputs via installations, our pathways illuminate many ways in which climate could affect installation readiness, as is discussed in more detail in the next section. Climate hazard information and the associated consequences could be incorporated via installation-level readiness reporting into DRRS relatively easily. DoD's *2017 Sustainable Ranges* report notes that the

> Army has implemented an Installation Status Report – Mission Capacity (ISR-MC) metric aimed at capturing the cost of repair to ranges and training lands due to extreme/atypical weather events. This metric was introduced in FY2016, and in FY2017 an additional metric will be implemented that aims to capture the impact of atypical weather events on loss of training days. Climate change is a long-term focus but the current approach is to capture the short-term atypical weather impacts with goals of identifying vulnerabilities and making sound decisions for future planning.[25]

---

[20] Department of the Navy, *Defense Readiness Reporting System–Navy (DRRS-N) Overview Course–PESTO*, CVN DRRS training facilitator's guide, undated; Department of the Navy, *Defense Readiness Reporting System–Navy*, OPNAV Instruction 2502.360A, October 17, 2014a.

[21] Department of the Army, *Installation Status Report Program*, Army Regulation 210–14, June 11, 2019, p. 1.

[22] Department of the Army, *FY18 ISR Infrastructure Implementing Instructions*, Version 1.0, October 1, 2017.

[23] Department of the Army, *Defense Readiness Reporting System—Army Procedures*, Pamphlet 220–1, November 16, 2011.

[24] DAF, *Force Readiness Reporting*, Air Force Instruction 10-201, December 22, 2020b, p. 34.

[25] U.S. Department of Defense, *2017 Sustainable Ranges*, Under Secretary of Defense (Personnel and Readiness), May 2017, p. 7.

While there is clearly a precedent for including climate-related information in Army and Air Force systems used for readiness reporting, it is unclear how consistently this information is captured, since a review of the data was beyond the scope of this research.

Strategic-Readiness Assessments

As with unit-level and installation-level reporting in DRRS, strategic-readiness assessments conducted as part of the JCCA are focused on the short term and do not systematically include assessments of weather or climate hazards. However, as of late, DoD has demonstrated increased focus on longer-term strategic-readiness reporting. At the time the research for this project was ongoing, it appeared that the Army was furthest along in developing a longer-term strategic-readiness assessment. The Army published Army Regulation 525-30, *Army Strategic and Operational Readiness*, in April 2020. The purpose was to describe the new Army Strategic Readiness Assessment (ASRA). This document defines *Army Strategic Readiness* as

> the Army's ability to provide adequate forces to meet the demands of the [National Military Strategy]. It is measured quarterly through the Army Strategic Readiness Assessment (ASRA) process . . . utilizing one Army and three Joint Staff mandated assessments to obtain an integrated view of current and future strategic readiness. Through Strategic Readiness Tenets (SRTs), the Army assesses leading and lagging measures and indicators to identify trends and risk in strategic and operational readiness across key Army resource areas. . . . ASRAs inform Army senior leaders of critical resource decisions necessary to address and mitigate shortfalls across the near-term (0–2 years), mid-term (2–5 years) and future (>5 years) time horizons.[26]

The Army's seven strategic-readiness tenets are manning, equipping, sustaining, training, leading, installations, and capacities and capabilities. According to Army Regulation 525-30, "Each readiness tenet contains measureable [sic] objectives and qualitative indicators which provide leading indicators of future changes in readiness."[27]

Although we were able to conduct only a limited review of information related to the ASRA and strategic-readiness tenets through our literature review and stakeholder discussions, the information we obtained indicated that the ASRA does not address potential future effects related to climate change, at least not specifically. This appears to be partly because any potential climate effects would have to be measurable and the Army has not yet identified the needed climate-related measures. This does not mean that such measures could not be identified going forward. The climate hazard pathways identified in our study outline the conceptual linkages from exposure to readiness effects. Modified DRRS reporting could provide some of the data needed to help mature the pathways into more-quantifiable measures, perhaps suitable for readiness forecasting, or at least "what if" analysis, and to help estimate how preparing for future climate impacts might require additional investments in adaptation that, in turn, might crowd out other demands for resources. In the remaining two sections of this chapter, we review additional potential climate integration points that may help inform the development and refinement of such quantitative measures. First, we discuss DCAT, which offers one quantitative approach to help assess potential future exposure to climate hazards, as discussed in more detail in the next section. Second, we discuss the Readiness Decision Impact Model (RDIM), the development of which could be used to help identify existing and needed quantitative inputs related to climate's effects on readiness.

---

[26] Department of the Army, *Army Strategic and Operational Readiness*, Army Regulation 525-30, April 9, 2020, p. 2.

[27] Department of the Army, 2020, p. 10.

Complementing Existing Exposure-Analysis Tools to Inform Readiness Exposure to Climate Risk

As noted in Chapter 2, DoD has already invested in analytic capabilities for hazard-exposure screening at installations by developing DCAT. DCAT is one example of an existing tool for measuring exposure to climate risk that could be expanded to incorporate readiness-specific effects, thereby informing the kinds of metrics that may be useful for readiness measurement and assessment. This potential climate integration point speaks to the Climate Adaptation Plan's "Climate Intelligence" focus area in that it provides decision-makers with metrics for understanding risk and threats to installations and the activities that are performed at these locations, although information learned from such a tool could in turn inform all three focus areas.

DoD already has plans to expand DCAT to include assessments of hazard sensitivity and adaptive capacity in the future. In its current form, DCAT is useful for understanding the overall level of exposure risk at installations, but it does not specify which impacts may occur or how readiness may be affected as a result of that exposure. Combining information from the climate hazard pathways with DCAT exposure projections may provide a more complete picture of future climate effects at installations than would be possible using either DCAT or the pathways alone. For example, analysts may use DCAT to identify the hazard with the highest projected exposure severity at an installation of interest, then refer to the pathways driven by that hazard to identify the specific readiness inputs most at risk at that installation.

Although not every pathway is relevant to installations, mapping the pathways that could occur at installations with high projected exposure to the associated hazards may enable a preliminary analysis of the variety of mechanisms by which readiness could be affected across installations. To illustrate this, we mapped the pathways from our climate assessment framework to hazard-exposure data in DCAT.[28] Figure 5.1 shows the total number of pathways from our dataset that could be of relevance at CONUS and rest-of-world installations and sites with high exposure to the relevant hazard under the assumption that installations or sites with high exposure to a hazard could experience readiness impacts related to that hazard exposure.[29] These results are based on DCAT climate projections by 2050 under a high-emissions scenario.

In addition, it may be useful to examine potential exposure on long-term time horizons, as indicated by multidecadal climate projections available in DCAT, at specific locations or theaters of interest where readiness inputs may be long lived. For example, investments in installation facilities often assume an infrastructure life cycle of 30 years or more, such that facilities built today—and the equipment, training, and force projection readiness inputs they support—will eventually be subject to levels of exposure associated with DCAT's midcentury time horizon (2050). Connecting information from the pathways with DCAT exposure data in this way remains primarily useful as a first-order screening method for understanding installation-level climate risk, and any adaptation or policy measures being considered in response to projected exposure should be based on more-granular information focused on specific exposure impacts below the level of the installation.

---

[28] Because our framework encompasses more hazards than are enumerated in DCAT, some of the pathways from our dataset were mapped to DCAT climate indicators instead. For instance, *cold* was mapped to the *Frost Days, Heating Degree Days,* and *5-Day Minimum Temperature* indicators in DCAT.

[29] An installation was defined as having *high exposure* if its DCAT WOWA score (or indicator score) for a given hazard was between the 75th and 99th percentiles relative to the other installations. In cases in which a hazard from our pathways was mapped to a DCAT indicator, we defined *high exposure* as the indicator value being between the 75th and 99th percentiles for at least one of the mapped indicators.

FIGURE 5.1

Number of Potential Pathways at CONUS and Rest-of-World Installations and Sites by 2050 Under a DCAT High-Emissions Scenario

Number of
Pathways at
Installation or
Site

- 0 to 20
- 20 to 40
- 40 to 60
- 60 to 80
- 80+

SOURCE: RAND analysis of high-exposure installations identified by DCAT.

Leveraging the Development of RDIM to Inform Assessment of Climate Risk

The last potential climate readiness integration point that we present in this chapter is an ambitious ongoing effort to build an enterprise-level readiness-prediction model. In this section, we specifically investigate how the climate readiness framework and associated pathways can inform the development of RDIM, a tool currently being built out by OSD(P&R). Opportunely, the timing of this study aligns with early phases of RDIM development, meaning that the climate readiness framework could be applied to identify ways to incorporate climate-related considerations throughout the development life cycle. At the time of writing, OSR(P&R) is considering expanding RDIM from a single illustrative model component to a more holistic model for predicting future joint force readiness under different high-level resourcing decisions. In this section, we describe a proof-of-concept application of the climate readiness framework to the development of one of RDIM's readiness-production "pipelines," laying the groundwork for the incorporation of climate risk based on the framework throughout RDIM's future implementation. Using the framework to inform the development of this production pipeline can help identify available and existing metrics for climate-related risk to readiness-generation processes, informing ways to incorporate climate into strategic, operational, and tactical decisionmaking. The continued application of the climate readiness framework as RDIM is built out could mean that this climate readiness integration point can eventually serve to incorporate climate into enterprise-level decisions about readiness resource allocation, allowing this integration point to speak to two of the Climate Adaptation Plan's focus areas for implementation.

Introducing RDIM

RDIM is a line of effort by OSD(P&R) that takes a system-dynamics approach to create models of the different readiness-production pipelines, and its planned objective is to support policy decisions by providing readiness projections given changes in key parameters, ultimately linking back to resourcing.[30] The initial proof-of-concept production pipeline implemented in RDIM leverages the aviation segment of an existing readiness-projection model from the U.S. Marine Corps.[31] Readiness is predicted in terms of the metrics used in current readiness reporting for personnel, equipment on hand, status of equipment, and training (i.e., P-, R-, S-, and T-ratings, which ultimately determine a unit's C-rating) by modeling the impact of policy decisions and readiness investments on key variables that influence unit-readiness metrics. Further development of RDIM aims to build out additional production pipelines that will eventually capture as many of the critical processes involved in generating joint readiness as possible, eventually yielding an enterprise-level view of the relationship between readiness resourcing decisions and readiness outcomes. If RDIM development is able to achieve this goal, and the climate readiness framework is used to inform development of each pipeline, RDIM could eventually capture the impacts of climate on resources needed for joint readiness.

Applying the Framework to Inform Development of an Existing Pipeline: Aviation Example

The proof-of-concept RDIM model logic is depicted in Figure 5.2, which shows the model for projecting the R-rating, or status of equipment, for aviation units in the U.S. Marine Corps. On the left side of the figure

---

[30] Deputy Assistant Secretary of Defense for Force Readiness and RDIM Modeling Team, discussions with the authors, February 3 and March 18, 2022; Meghann Myers, "New Data Model Predicts How Deployments Affect Future Readiness," *Air Force Times*, December 29, 2022.

[31] Headquarters Marine Corps Plans, Policies, and Operations staff, communications with the authors, October 2021.

FIGURE 5.2

Example Climate Narrative Annotation of Aviation Pipeline

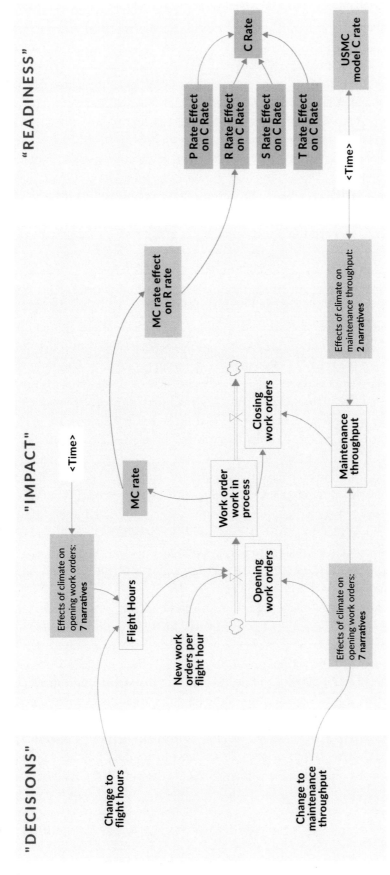

SOURCE: RAND annotation of figure based on Deputy Assistant Secretary of Defense for Force Readiness and RDIM Modeling Team, email communications and discussions with the authors, February 3 and March 19, 2022.

are the *decisions*, which are essentially policy levers that are parameterized to affect the key variables of the model. The decisions are a change to flight hours and a change to maintenance throughput. *Flight hours* represents how much aircraft are flown, the logic being that the more aircraft are flown, the more often they will require maintenance and reduce their mission-capable (MC) status and their unit's R-rating. The relationship between flight hours and the opening of work orders is based on statistical relationships of historical data. The maintenance throughput affects the speed at which aircraft are maintained or repaired when a work order is opened; a faster throughput would return a non-MC (NMC) aircraft back to MC status quickly and keep the R-rating high.[32]

The RDIM proof of concept was originally designed to model the relationship between flight hours, work orders, and maintenance throughput. The purple boxes in Figure 5.2 illustrate where climate change would create vulnerabilities in the modeled aviation pipeline for status of equipment. We identified eight climate hazard pathways (representing seven distinct narratives) that would have an impact on flight hours, seven climate hazard pathways (representing two narratives) that would affect maintenance throughput, and 14 pathways (representing seven narratives) that would affect the rate at which work orders are opened. In our annotation, note that we consider aviation generally across DoD and not specifically for the U.S. Marine Corps.

In this section, we walk through the set of climate hazard pathways that could affect flight hours, corresponding to the topmost purple box. Annotation related to the other two purple boxes is detailed in Appendix C. Figure 5.3 shows the seven narratives that were identified as potentially affecting flight hours. The narratives are associated with a variety of hazards and readiness impacts (see the "Impact" column in the figure and the list of climate hazard pathways in the separate online appendix[33]), which relate to a variety of readiness inputs (see the colored and labeled arrows in the figure). The variety of hazards, impacts, and inputs is consistent with the fact that aircraft are used for training and operations but also play a role in force projection. In the descriptions of each narrative, one can see that the relevant pathways are ones where an aircraft would have reduced performance, reduced ability to perform a mission, reduced ability to train, and increased threat of safety issues.

In Figure 5.4, we further characterize the seven narratives associated with impact on flight hours. Starting at the left, the colored arrows show that different hazards produce different types of effects, with recurrent nuisance flooding and heat causing primarily long-term "stressor" effects and wind and wildfire causing primarily short-term "shock" effects related to catastrophic events. Both the hazard and the type of effect are important when considering what sorts of indicators to use to quantify the relevant climate risk for incorporation into a model. Different metrics may capture the threat from shocks versus stressors, and different metrics may be required to measure the possible effects on aircraft.

In addition to differences in the type of climate impact on the production pipeline, we can characterize different kinds of effects on flight hours. Figure 5.4 marks each narrative in terms of whether the impact of the pathway is likely to be reduced quality of flight hours or reduced ability to execute flight hours. Six of the seven relevant narratives would likely lead to a reduced quality of flight hours in the sense that the aircraft may experience a deviation from its original flight plan due to unsuitable conditions. These types of effects would ultimately affect the mission or training of the intended flight, which is important. However, these effects are not necessarily relevant to the status-of-equipment (R-rating) pipeline; capturing this kind of impact may need to be noted for incorporation as RDIM is built out to include training- and force projection–related pipelines. Four climate narratives are likely to lead to a reduced ability to execute flight hours, such as fully canceling a sortie or blocking the ability to execute the intended mission or training. In

---

[32] Deputy Assistant Secretary of Defense for Force Readiness and RDIM Modeling Team, email communications and discussions with the authors, February 3 and March 18, 2022.

[33] Best et al., 2023.

FIGURE 5.3

Climate Narratives Related to Flight Hours

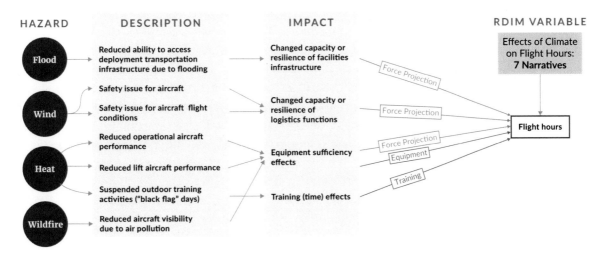

SOURCE: Features information from Deputy Assistant Secretary of Defense for Force Readiness and RDIM Modeling Team, email communications and discussions with the authors, February 3 and March 19, 2022.

FIGURE 5.4

Characteristics of Climate Narratives Related to Flight Hours

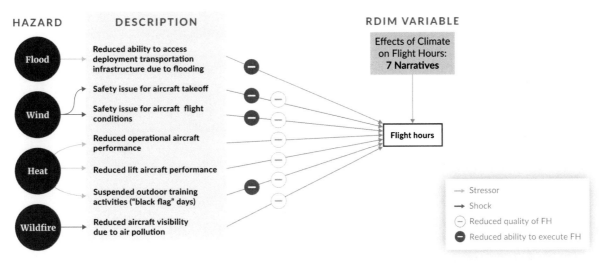

SOURCE: Features information from Deputy Assistant Secretary of Defense for Force Readiness and RDIM Modeling Team, email communications and discussions with the authors, February 3 and March 19, 2022.

this pipeline's logic, that means fewer opportunities for the aircraft to require maintenance and lose its MC status—a good thing for the R-rating. However, if we consider readiness holistically, to include the T-rating or ability to execute a mission, these narratives will have negative effects on readiness, again leaving the effect to be incorporated in future production pipelines.[34]

As mentioned before, one benefit of the readiness pipeline annotation exercise described above is that it allows us to identify *where* in the pipeline logic climate change could inject risk. Another benefit is that this

---

[34] The climate hazard pathways for this project focused explicitly on negative climate effects, and the potentially positive effect on the R-rating is incidental and may be outweighed by the negative impact on other readiness inputs.

exercise can highlight what data and indicators could be used to track and alert decisionmakers to the risk. In Figure 5.5, we highlight potential data and indicators for the seven climate narratives relevant to flight hours.[35] This identification of datasets is illustrative and is not intended to suggest that these are the best or only datasets relevant for this pipeline. One indicator, WXX (weather) deviations, shown with the green arrows in the figure, relates to the cause of the impacts on readiness. A weather deviation is recorded in the data whenever there is a change to an original flight plan due to a weather event. Ideally, for tracking climate risks, there would be metadata or further specification as to the type of weather event that caused the disruption, allowing us to better connect the indicator to a particular hazard and pathway. The other data and indicators identified relate more to the realized effects of the climate pathways and are noted in purple, yellow, and red in the figure. We provide a potential source for the data and indicators in Table 5.1, using U.S. Air Force data as an example. Starting with the installation-readiness data (purple arrow), among the many factors tracked for DoD installations, a reduced ability to access deployment transportation infrastructure due to flooding would likely be reported. For the DAF, this might be captured in installation reports in DRRS-S and the Installation Health Assessments. The other narratives that relate to safety issues, reduced performance, or suspended training all potentially reduce flight hours. Data related to these narratives could include hours flown or UTE (aircraft utilization rate) cancellations, both of which are tracked in a DAF database called Logistics, Installations and Mission Support-Enterprise View (LIMS-EV). Lastly, the pathways related to aircraft performance may be informed by some type of maintenance data, such as depot rates or status and MC and NMC status and rates as indicators. For the DAF, these metrics are currently tracked in LIMS-EV.

Incorporation of Climate Risk into Future RDIM Development

With the additions suggested in the previous section, the single-pipeline, proof-of-concept RDIM could already be used to investigate the extent to which weather deviations or black flag days affect R-ratings for

FIGURE 5.5

Data and Indicators for Climate Narratives Related to Flight Hours

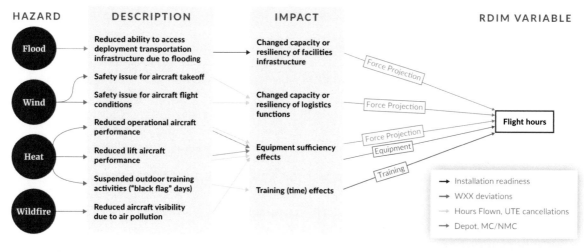

NOTE: UTE = aircraft utilization rate; WXX = weather.

---

[35] This analysis was conducted by examining aviation-related data from the U.S. Air Force; however, there are equivalent data sources in the other services that have their own aviation platforms.

TABLE 5.1

Data and Indicators for Climate Narratives Related to Flight Hours

Data or Indicator	Potential Source
Installation readiness	DRRS-S reports from installations; Installation Health Assessments
Weather deviations	LIMS-EV
Hours flown, UTE cancellations	LIMS-EV
Depot, MC/NMC	LIMS-EV

NOTE: UTE = aircraft utilization rate. For more on Installation Health Assessments, see Joe Bela, "AFIMSC Analytics Tool Expands, Helps Leaders Prioritize Investments," press release, May 28, 2020.

aviation. In combination with geographical filtering of data to align with specific climate hazard exposure, additional types of questions could be asked, such as the following:

- Which climate risk indicators should be watched by location?
- What regions or installations are likely to have an R-rating impact from climate change?
- Which climate adaptation investments would most reduce R-rating impacts?

Furthermore, as RDIM is expanded to include other production pipelines that predict changes in P-ratings, T-ratings, and S-ratings, additional climate-related readiness questions could be answered, eventually allowing for more-comprehensive consideration of climate change risk across the readiness enterprise.

Drawing on previous RAND work, we developed potential aviation production pipelines for T-ratings and P-ratings. Here, we walk through the T-rating production pipeline shown in Figure 5.6, which is considerably more complex than the R-rating example.[36] Starting with "Decisions" in the left of the figure, there are still the change to flight hours and the change to maintenance throughput from the R-rating pipeline. We added *change to personnel availability (to train)* and *change to simulator hours* to incorporate personnel- and training-system decision levers, respectively. The personnel availability, flight hours, and maintenance throughput are meant to account for the live piece of aviation training, and the simulator hours and personnel availability account for aviation simulator training. Personnel availability to train can be influenced by many factors, including deploy-to-dwell ratios, training requirements, exercises, ground training, office duty, simulator days, professional military education, no-fly training days, leave, and medical status. Ultimately, the production of T-ratings depends on the completion of training requirements, so the main modeling effort would be to link personnel availability, simulator hours, flight hours, and maintenance throughput (i.e., aircraft generation) to training completion. Arrows between pieces in the T-rating logic are colored to show whether the modeling between quantities is a simple calculation based on established business rules (gray) or something that will be modeled using system dynamics or historical statistics (orange).[37] As one can imagine by comparing this potential T-rating pipeline and the R-rating pipeline, there are overlapping variables and interdependencies. For instance, the climate pathways that would decrease flight hours from the R-rating pipeline example would also take away flight hours that contribute to fulfilling training requirements and negatively affect the T-rating. As RDIM continues to be developed, there will be more opportunities to examine where additional climate pathways have relevance, identify indicators and data, and investigate more-sophisticated readiness questions that consider interrelationships between readiness inputs.

---

[36] T-rating production pipeline logic is influenced by previous work, particularly Dara Gold, Bart E. Bennett, Bradley DeBlois, Ronald G. McGarvey, and Anna Jean Wirth, *A Modeling Framework for Optimizing U.S. Air Force Fighter Pilot Access to Advanced Training Ranges*, RAND Corporation, TL-A169-1, 2020.

[37] See Appendix C for the potential aviation personnel readiness-production pipeline.

FIGURE 5.6

Potential Aviation Training Readiness-Production Pipeline

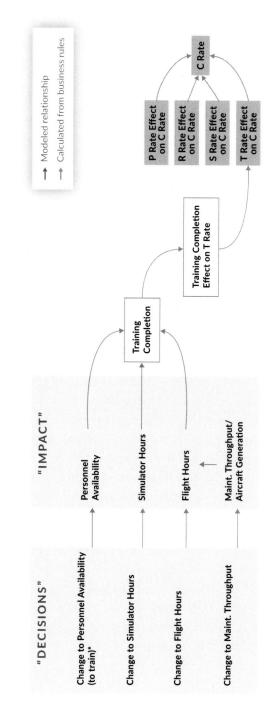

SOURCE: Features information from Deputy Assistant Secretary of Defense for Force Readiness and RDIM Modeling Team, email communications and discussions with the authors, February 3 and March 19, 2022.

From Climate Readiness Integration Points to Implementation

The potential climate readiness integration points presented in this chapter are mere examples of the ways that climate risk could be integrated into existing DoD and service processes. However, even this set of examples speaks to each of the three focus areas for "Implementing Climate-Informed Decision-Making" laid out by the DoD Climate Adaptation Plan.[38] Development and application of the various processes and models discussed in this chapter is already underway. These examples show how the framework might be used to make changes—of varying size and cost—to these processes in an effort to integrate climate risk, potentially helping inform decisions about where such integration is feasible, useful, and cost-effective.

---

[38] U.S. Department of Defense, 2021b, p. 6.

CHAPTER 6

Findings and Recommendations

In this chapter, we revisit and organize the main findings from our literature review, outreach to climate and readiness stakeholders, and development and application of the climate readiness framework. The findings summarize important takeaways on current linkages between climate and readiness across DoD and service communities, the possible current and future impacts of climate hazards on readiness, and the implications for integrating climate more fully into readiness decisionmaking. We then build upon these findings to offer recommendations for strengthening linkages between climate and readiness strategy development, planning and assessment, risk management, and decisionmaking within DoD specifically and for other stakeholders at the intersection of climate and national security more generally. The recommendations cover ways to continue to refine the climate readiness framework, as well as ways to apply the framework and further incorporate climate into readiness reporting, readiness assessment, analytic capabilities, and decision processes.

Findings: The Link Between Climate and Readiness

From our literature review, conversations with stakeholders, and review of how climate informs current processes and models, we found that, in general, climate and readiness thinking are not integrated in DoD and the services. In this section, we briefly revisit some of the key takeaways from Chapters 2 and 5 that speak to the current links between readiness and climate thinking, as well as opportunities to build out those links.

Climate and Readiness Are Understood and Studied by Separate Communities

Throughout our discussions with readiness experts in DoD and the services, we found that climate effects, especially over the long term, are not part of the readiness dialogue. Although DoD is considering ways to improve climate literacy, that literacy is not yet common in the readiness community. Readiness operators are focused on short-term weather effects and are not yet fully considering the effects of future weather on future readiness. Furthermore, the readiness operators are not the same individuals as the nonoperational managers—i.e., installation commanders, regional commanders, and higher headquarters—who consider longer-term, strategic questions that are important to supporting the readiness enterprise. Connective tissue going the other way—from climate thinking to readiness thinking—shows somewhat more of a linkage, with readiness being identified throughout our discussions with climate thinkers in the national security space as one of the processes that might be disrupted by climate. However, much of DoD's work related to climate is still being developed and is not yet providing immediate and specific inputs to the readiness community. For example, DCAT assesses exposure to various climate hazards at installations across the services, but its current instantiation supports primarily departmentwide hazard-exposure screening rather than installation-level operational risk management or investment planning. Below, we identify several potential reasons for the disconnect between climate and readiness.

Climate and Readiness Thinking Operate on Different Timescales

The relevant time dimensions for considering climate and readiness generation are different. Readiness leaders have immediate demands that often preclude them from dedicating attention to potential challenges that do not have short-term impacts to operational and tactical outcomes, and longer-scale strategic planning may be the responsibility of other leaders. Other than immediately observable impacts of specific "now" weather events, climate effects on readiness need to be assessed over a much longer time frame. Time horizons for DoD and service climate-related strategies, as well as time horizons for DCAT projections, look out to 2030 and beyond, while departmentwide readiness plans, such as the Global Force Management Allocation Plan, look out no more than a few years.[1] Unit commanders lack direct responsibility or control over anything other than immediate, short-term adaptation to weather effects, with broader, longer-term, far-reaching climate responsibilities falling to nonoperational leaders—i.e., installation commanders, regional commanders, and higher headquarters. Climate will, however, have significant impact on the "near-term readiness" of the future; as climate shifts, the kinds of events that require attention from commanders could change in scale, frequency, and nature. The necessary planning and awareness to account for impacts on future readiness might not yet be a priority within the readiness community.

Climate and Readiness Operate at Different Levels

Commensurate with this difference in timescales, many readiness-related decisions occur at the operational and tactical levels, but climate-related decisionmaking will need to occur primarily, although not only, at the strategic level. At the strategic level, responsibilities for decisionmaking and planning at the intersection of climate and readiness are not always clear. Most readiness experts with whom we spoke noted that the issues were "above their pay grade" or generally required an integrated approach that extended beyond their direct areas of responsibility. Experts from both the readiness and climate communities said that climate was only starting to be included in broader, long-range planning by DoD and the services, noting that individuals with climate expertise are not typically present in decision forums and that senior decisionmakers are not always being educated to present climate issues from a strategic and all-encompassing perspective. It is not clear whether the existing organizational structure enables communication of strategic climate-related matters down to the operational and tactical levels in an actionable way, or whether tactical and operational decisionmakers will have the needed information and purview to prepare for climate change effects over the longer term. Importantly, climate will have a future impact on short-term operational readiness, through future weather events, implying that these issues should be of interest to the readiness community across levels.

Short-Term Climate Impacts May Be Small and Locally Mitigable

Although the disconnect described in this section may seem disheartening to some audiences, our outreach confirmed that there may be a very good, real-world reason why connective tissue has not yet been built between climate and readiness: Climate impacts that are likely to manifest on the timescales of interest to the readiness community may be both small and manageable. Current and short-term impacts from climate may be locally addressed using the set of tools currently available to operational and tactical decisionmakers. In today's environment, it remains possible to view climate change mainly as a "training condition" related to short-term weather events. However, experts with whom we spoke consistently noted that military units are now "operating closer to the margin" as the cumulative impacts of readiness impediments are being realized. Such impacts include many external factors that are unrelated to climate, but also lost training days due to local weather. As the "margin" decreases, the impact may grow and the ability to adapt locally may decline. While there is awareness that the services are moving closer to the thresholds at which cumulative impacts,

---

[1] See, e.g., U.S. Army War College, *How the Army Runs: A Senior Leader Reference Handbook*, January 29, 2020, pp. 2-4–2-7.

possibly across multiple hazards, will cause significant effects or become excessively costly, these thresholds are not well understood.

Climate Risk Is Not Integrated with Readiness Measurement or Reporting

In addition to not being a central theme in readiness discussions, climate-related effects are not well integrated with readiness reporting and assessment. DoD and the services have robust and well-established processes for measuring, recording, and assessing readiness from the tactical level through the strategic level. The primary assessment processes—unit-readiness reporting and strategic-readiness assessment—are focused on short-term readiness and do not consider the potential readiness impacts of factors that may play out over the course of years or decades, such as climate change. Climate-related effects would likely be tracked via weather-related metrics, but current tracking of weather events is not intended to inform awareness of broader climate trends. Metrics are designed for the primary purpose of reporting upward about the current and short-term state of force readiness. They are intended to be snapshots in time, so determining trends or cumulative effects, such as those associated with climate change, is not fully supported. This has several implications that make the metrics less suited to track or assess climate impacts. First, metrics and collected data are generally focused on outputs; causal information, about whether disruptions or reduced readiness occurred because of a given factor, such as weather, is generally not tracked in a structured format, if it is tracked at all. Second, the current fidelity of readiness measures makes it difficult to see the impacts of weather on readiness, because each individual impact is perceived by readiness reporters as a relatively minor current effect caused by weather. Third, the net results of actions taken by commanders to overcome climate issues are not recorded and therefore cannot be aggregated. Adaptations to the current levels of weather-related impacts tend to be local, with individual commanders overcoming specific concerns or disruptions, meaning that these adaptations are not reflected in a reporting mechanism designed to report issues upward.

Current Processes, Tools, Models, and Assessments Show Promise as Points of Future Integration

While the findings above describe shortcomings of the current integration of climate risk and readiness decisionmaking and assessment, we found many potential ways that DoD and the services could use existing mechanisms to strengthen the connection between these two domains. The DoD Climate Adaptation Plan stresses the importance of actually "Implementing Climate-Informed Decision-Making," highlighting focus areas related to the incorporation of "Climate Intelligence" into all types of threat assessment, the incorporation of climate into "Strategic, Operational, and Tactical Decision-Making," and the incorporation of climate into "Business Enterprise Decision-Making."[2] Our literature review and outreach identified promising potential integration points between climate and readiness, detailed in Chapter 5, that speak to all three of these focus areas. Recommendations for next steps for some of these integration points are provided in a later section.

Findings: Organizing and Prioritizing Climate Impacts on Readiness

Our analysis demonstrates several ways in which the climate readiness framework can help illuminate readiness vulnerability to climate risk. In this section, we outline key takeaways from Chapters 3 and 4 that speak to the contributions of the climate readiness framework and pathways.

---

[2] U.S. Department of Defense, 2021b, p. 6.

The Climate Readiness Framework Can Provide a Useful Structure for Understanding Readiness Vulnerability to Climate Risk

The framework provides a comprehensive and standardized means to organize the ways in which climate change could affect readiness and to subsequently visualize and analyze patterns among elements in the framework. These patterns can reveal areas of risk that may then be targeted for further investigation or prioritized for adaptive action, or both. For example, areas of the framework where numerous connections exist between multiple hazards and multiple readiness inputs, particularly where such connections implicate multiple types of impacts, can help highlight areas of the readiness system that are vulnerable in a variety of potentially overlapping ways. Conversely, areas of the framework with fewer connections suggest a narrower scope of impact on readiness for those pathways, but they could also indicate where more research is needed. Visualization techniques for the framework and pathways can facilitate pattern identification.

The Full Impact of Climate Change on Readiness Requires Consideration of All Climate Hazard Pathways in Aggregate

As our stakeholder engagements revealed, many climate impacts on readiness are currently seen as minor because of a perception that they may be temporally or spatially distant, easily mitigated through adaptation, or both. However, while minor hazards in isolation may be overcome relatively easily, a confluence of multiple pathways simultaneously could quickly aggregate to mission-level impacts in death-by-a-thousand-cuts fashion. Additionally, our conversations frequently included questions about the "most important" climate effect or the "biggest" hazard, but a review of the climate readiness framework and set of pathways shows that many hazards, through many mechanisms, could have similar impacts, leading to an aggregate effect on readiness inputs. Prioritizing these impacts and the kinds of disruptions that could occur through the combined effects of multiple hazards and mechanisms may be more fruitful than attempting to prioritize the hazards themselves. Relatedly, even small hazard perturbations could result in significant impacts if readiness is already at the margins. In this way, even low hazard exposure can reduce the "readiness headroom," straining resources and leaving decisionmakers with less margin for error.

Patterns Among Pathways Can Be a First Step Toward Prioritization

Together, quantitative analysis of the frequency of and connections between elements of pathways and qualitative analysis of their relative sensitivity and adaptive capacity can inform preliminary decisionmaking regarding where and how to prioritize action to reduce impacts of climate change on readiness. The frequency of a hazard within the pathways should be understood as an indicator of the number of "opportunities" the hazard may have to affect readiness rather than as the sole determinant of priority. Similarly, focusing the frequency analysis on readiness inputs highlights those elements of the readiness system that have exposure to the widest variety of hazards and impacts and, therefore, the greatest number of potential avenues for disruption, many of which might not be apparent at the tactical, local, or individual unit level. Readiness inputs that face impacts from multiple hazards, and multiple impact types across hazards, are more at risk if exposure events occur concurrently or if the time between events is insufficient for a full recovery. The initial indicators of a hazard's (or readiness input's) prevalence among the variety of pathways identified should form the basis for additional analysis to better understand the implications of exposure at specific geographic locations where the hazard or readiness input may be especially common or critical.

Information Beyond Hazard Representation in Pathways Is Required to Understand the Vulnerability of Readiness to Climate

Understanding a readiness input's vulnerability to climate change requires consideration of the input's adaptive capacity and sensitivity to a variety of hazards. We performed a modified Delphi exercise to qualitatively assess the vulnerability of the various pathways to climate hazards. We found that the Delphi exercise is a useful, structured approach for generating a high-level characterization of the narratives or pathways. It can be performed relatively quickly, at a level of analysis that is suited to the decisions at hand and the amount of available information. Furthermore, the richness of the discussion around the ratings, such as the factors that participants considered, and the determinants of their assessment ratings can increase DoD's awareness and management of climate risks to readiness. For example, the approach can be used to delineate the key assumptions among participants about the drivers that influence a climate hazard's potential to affect a readiness input and to improve understanding of the vulnerabilities of a readiness input to those hazards. A lack of consensus can reveal key factors that shape the vulnerability of a readiness input to the climate hazard. It can inform a discussion of what adaptation options exist; what resources need to be added, including understanding the importance of system slack or redundancy; and how the various functions or stakeholders need to consider climate hazards. The discussions can also reveal where more information is needed and should be gathered. Furthermore, the framework, visualizations, and Delphi exercise can be used to engage subject-matter experts and stakeholders to raise awareness of potential risks and vulnerabilities and to prioritize investments and action, including identifying system indicators that should be monitored.

Our Delphi exercise revealed that the readiness inputs of *people*, *force projection* (relating to transportation infrastructure), and *equipment* (relating to logistics functions) were rated as having high adaptive capacity, as they were assessed as having several options for reducing harm or making temporary adjustments to climate impacts. In contrast, *force projection* (relating to facilities) and *equipment* (relating to sufficiency) were assessed as having low adaptive capacity, owing to having relatively few inexpensive options for hazard mitigation. *Force projection* (relating to demand for DoD capabilities), *force projection* (relating to energy and water), and *equipment* (relating to logistics) were rated as relatively insensitive to climate hazards, while *force projection* (relating to facilities), *equipment* (relating to facilities), and *equipment* (relating to efficacy) were rated as relatively sensitive. Overall, these results suggest that inputs relating to facilities and large infrastructure and equipment investments may be the most vulnerable to future climate change, as these inputs are often tied to design standards and performance requirements that might not account for changing hazards and are often expensive to replace.

Recommendations: Future Links Between Climate and Readiness

The preceding sections outlined key findings from our literature review, stakeholder engagements, application of the climate readiness framework, pathways dataset and accompanying visualizations, and potential integration points with existing DoD processes and tools. Here, we offer several recommendations that build upon these findings in order to further develop and apply the climate readiness framework and, ultimately, further integrate climate into readiness decisionmaking.

The Department of Defense Should Consider Developing, Adapting, and Building Upon a Climate Readiness Framework

Our study demonstrates some of the benefits of a high-level, comprehensive, and integrated framework for organizing and finding patterns among the many ways that climate change may affect readiness. To realize some of these benefits, DoD should consider developing and adapting the framework presented in this report or developing a similar framework that could help reveal insights about DoD and service readiness vulnerabilities to climate risk. A climate readiness framework can assist in

- identifying causal pathways between climate hazards and readiness outcomes
- revealing patterns among pathways and highlighting common hazards, impacts, and readiness inputs, as well as areas of limited awareness of climate impacts on readiness
- revealing qualitative differences between types of pathways, such as differentiation between stressors and shocks or differentiation between readiness inputs (such as training) that may be more sensitive to transient events and readiness inputs (such as equipment) that may face more-protracted climate exposure over a period of decades
- communicating the additive nature of readiness vulnerabilities across hazards, including building an understanding of pathway prevalence not only by hazard but also by second-order effect, by impact, and by readiness input
- supplementing ongoing data-driven efforts that focus on identification and development of metrics.

Once such a framework is established, DoD could supplement it with information on sensitivity and adaptive capacity of readiness inputs, allowing for prioritization of areas of high concern across hazards, impacts, and readiness inputs. The illustrative Delphi exercise outlined in Chapter 4 suggests one potential avenue for collecting such information. Delphi analysis could include analysts and decisionmakers responsible for generating readiness inputs who may be able to assist with selection of plausible indicators for monitoring. Prioritization methods, such as quantitative frequency analysis and qualitative Delphi elicitation, should consider the combined effect of numerous climate hazards in aggregate rather than focus only on hazards represented frequently in pathways. Regional case studies could also help reveal areas of high vulnerability and high exposure of readiness inputs concentrated in critical locations.

Other uses of the framework could be to explore the readiness impacts of different scenarios, such as simultaneously occurring hazards across the eastern seaboard, or to examine how climate hazards could have cascading effects throughout the system. The framework could also be used to build awareness and climate literacy in different training events or forums.

The Department of Defense Should Continue to Strengthen Linkages Between Climate and Readiness

DoD should find ways to integrate climate and readiness by bringing together climate and readiness experts and providing accessible and authoritative climate data relevant to readiness decisionmaking within DoD and the national security space. More specifically, DoD should consider doing the following:

- Ensure that climate experts are present at high-level strategic-readiness decisionmaking forums to help ensure that climate threats to readiness are consistently treated and that cost-effective means to mitigate these risks appropriate to the levels of uncertainty are identified.
- Continue efforts to improve climate literacy and awareness appropriate for different functional areas, as emphasized in the DoD Climate Adaptation Plan, within the readiness community so that climate risks can be properly identified and managed. This should include providing data and decision-support tools that are easy to use at the operational level. Existing materials, such as AHTAs and service-level installation resilience plans, could provide starting points for identification and collection of needed authoritative data.
- Ensure that there is a clear organizational structure and delineation of responsibility for identifying and addressing climate-related risks at the tactical, operational, and strategic levels. Strategic decisionmak-

ing related to, for example, installations is already using and integrating climate-related information at least to some extent, but it might not be focused on readiness impacts. It is not clear whether the existing organizational structure enables communication of strategic climate-related matters down to the operational and tactical levels in an actionable way, or whether tactical and operational decision-makers will have the needed information and purview to prepare for climate change effects over the longer term. Climate will have future tactical- and operational-level impacts through the mechanisms of future weather, and it is unclear whose responsibility it is to understand and manage related risks.

- Build on existing strategic guidance on climate and climate change, such as DoD Directive 4715.21, to ensure that decisionmakers have the correct training, data, decision-support tools, and access to needed expertise to implement strategies in a practical manner while continuing to perform primary missions.

Readiness Reporting Should Be Structured to Provide Information to Track Climate Impacts

DoD should find ways to integrate climate and readiness by building climate-related information into readiness reporting and assessment within DoD and the national security space. More specifically, DoD should consider

- using a framework to help identify existing data that could capture climate risk to and effects on readiness and facilitate reporting and modeling, as demonstrated through the RDIM annotation exercise in Chapter 5
- using a framework, such as the one presented in this report, to identify gaps in existing data
- adding causally focused metrics to existing output-focused readiness metrics in order to capture weather-related causes for training disruption in a structured format, as well as facilitate readiness management
- adding adaptation-focused metrics that capture small, locally administered efforts, thus facilitating upward visibility of the cumulative effects of small disruptions and overall trends related to shrinking "readiness headroom"
- including climate more explicitly in MA and hazard threat-assessment constructs, facilitating climate-related assessments and data collection, and providing guidance and resources to ensure that planners have the best available information to avoid ineffective or counterproductive investments
- including climate more explicitly in semiannual readiness reports to Congress or similar high-level products.

The Department of Defense Should Continue to Identify and Prioritize Opportunities for Implementing Climate-Informed Decisionmaking

Our study demonstrates that many current processes, tools, models, and assessments show promise as future integration points between climate and readiness. Continued identification of integration points that speak to the focus areas identified in the Climate Adaptation Plan, such as "Implementing Climate-Informed Decision-Making,"[3] could allow the development and revision of such processes and tools to account for climate-related risk without a need to develop new and parallel processes. Potential integration points identified in Chapter 5 of this report should be considered, prioritized, and built out where applicable. Consideration of climate risk throughout the development of RDIM may be a promising test case for integration of climate risk during development of a DoD-wide readiness-focused modeling capability.

---

[3] U.S. Department of Defense, 2021b, p. 6.

Conclusion

Military readiness is a priority for DoD and the services. The DoD Climate Adaptation Plan includes lines of effort toward maintaining installation and personnel readiness in the face of a changing climate, and the 2021 National Intelligence Estimate concludes that climate change has the potential to moderately strain military readiness by 2040.[4] We hope that the three primary outputs of our study provide a useful addition to the ongoing and growing dialogue related to climate change and readiness in DoD and the services. The climate readiness framework presented in this report is a high-level and comprehensive system for organizing the connections between climate conditions and joint force readiness and can help identify and characterize possible climate effects on readiness, both now and in the future, as decisionmaking becomes increasingly informed by projections of future climate change. The set of pathways collected throughout the study is a starting point for listing the mechanisms through which climate hazards have an opportunity to affect readiness inputs and represents a first step toward evaluating climate risks to readiness. The preliminary exploration of patterns and addition of information on sensitivity and adaptive capacity show how the pathways and framework could be used to begin prioritizing areas of high concern, developing indicators to monitor, or identifying areas requiring adaptation investments. Finally, the set of identified climate hazard integration points presents both a potential menu for short-term integration of climate into existing processes, tools, models, and assessments and a starting point for identification of additional opportunities. DoD is making great strides in terms of developing strategies and high-level information sources related to climate impacts. We hope some of the findings and recommendations in this report can assist with moving toward implementation that supports real-world decisionmakers.

---

[4] National Intelligence Council, 2021.

Climate Hazard Pathway Network Diagrams

This appendix provides network diagrams for additional hazards considered in the study. Seven diagrams (Figures A.1–A.7) respectively depict the sets of pathways associated with storms; inland and coastal flooding; wildfire; drought; land degradation; cold; and ocean anomalies, wind, sea ice retreat, and non-flooding precipitation. An eighth diagram illustrates pathways associated with all hazards together (Figure A.8). These network diagrams are analogous to Figure 4.2 and identify the climate hazards, associated second-order effects, readiness impacts, and readiness inputs for pathways associated with the relevant hazards.

FIGURE A.1
Storm Pathways Network Diagram

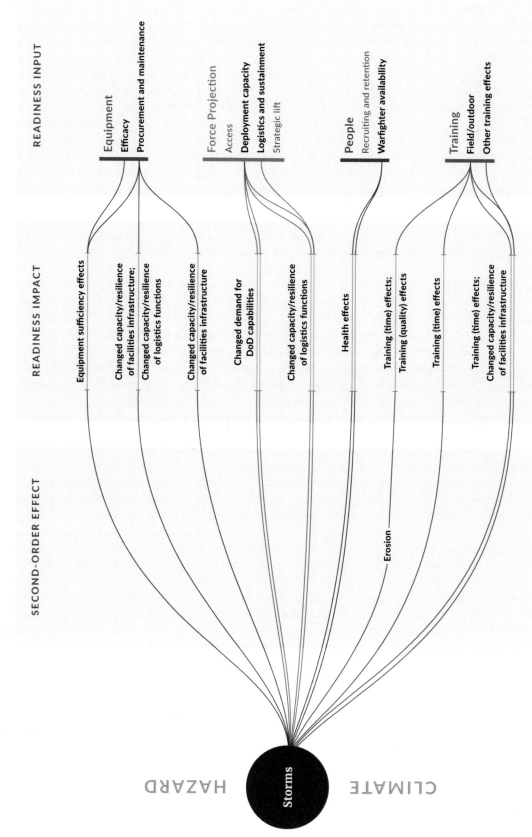

FIGURE A.2
Flooding Pathways Network Diagram

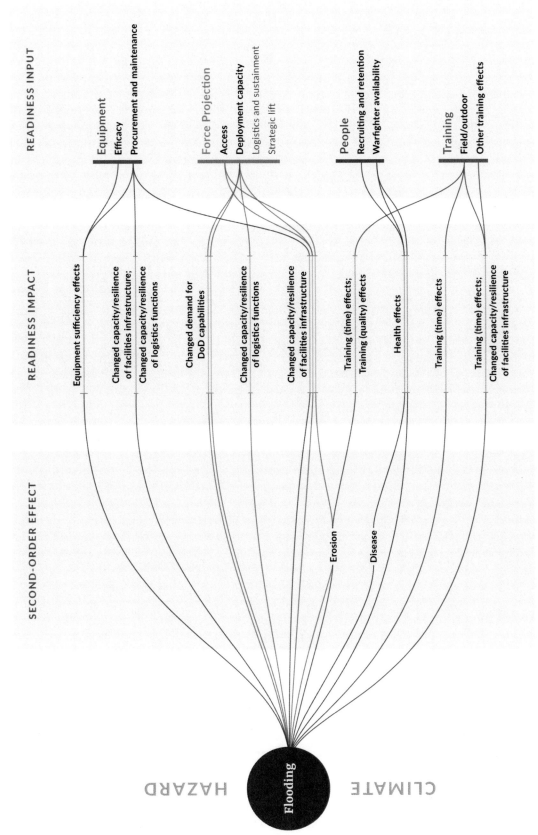

FIGURE A.3
Wildfire Pathways Network Diagram

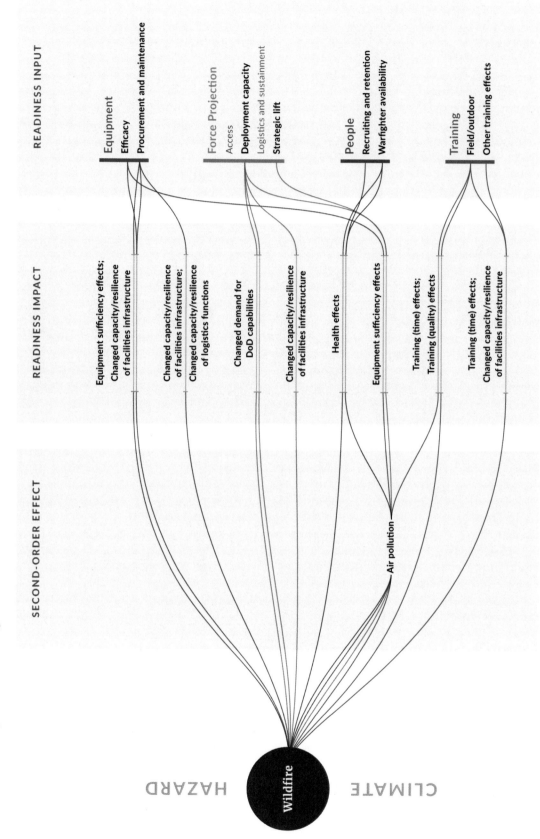

FIGURE A.4
Drought Pathways Network Diagram

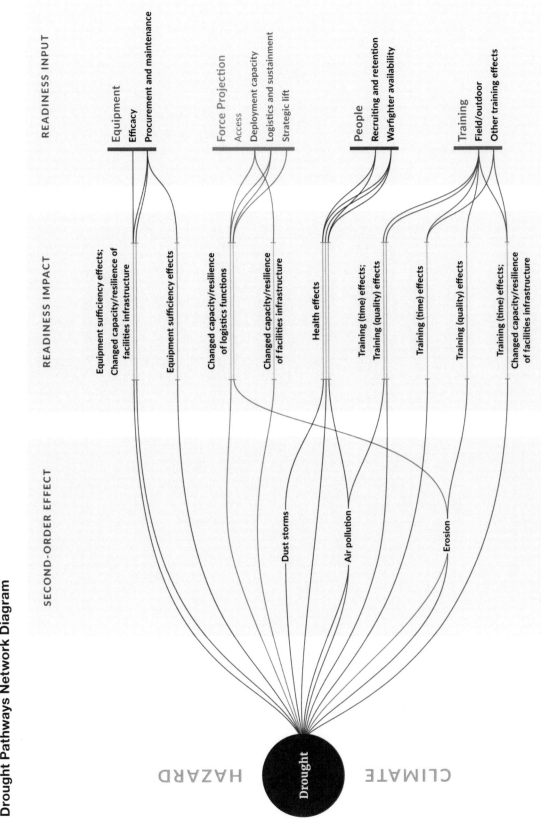

FIGURE A.5
Land Degradation Pathways Network Diagram

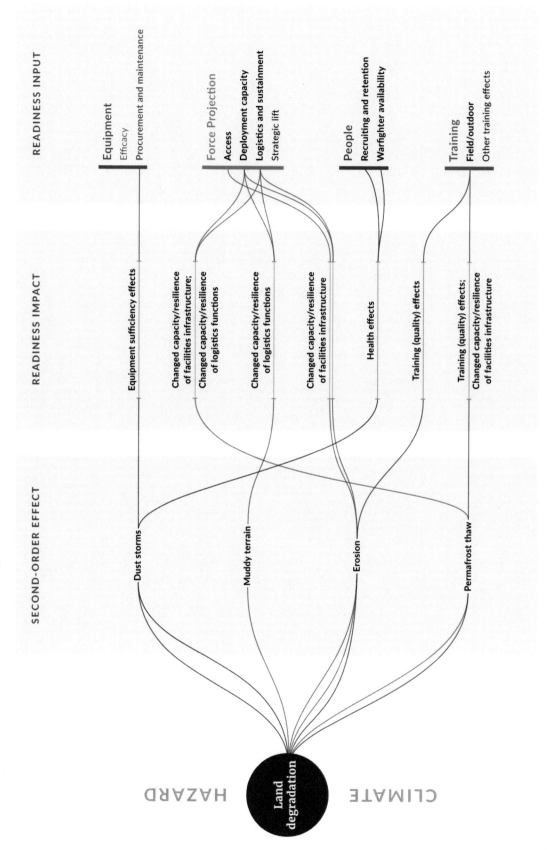

FIGURE A.6
Cold Pathways Network Diagram

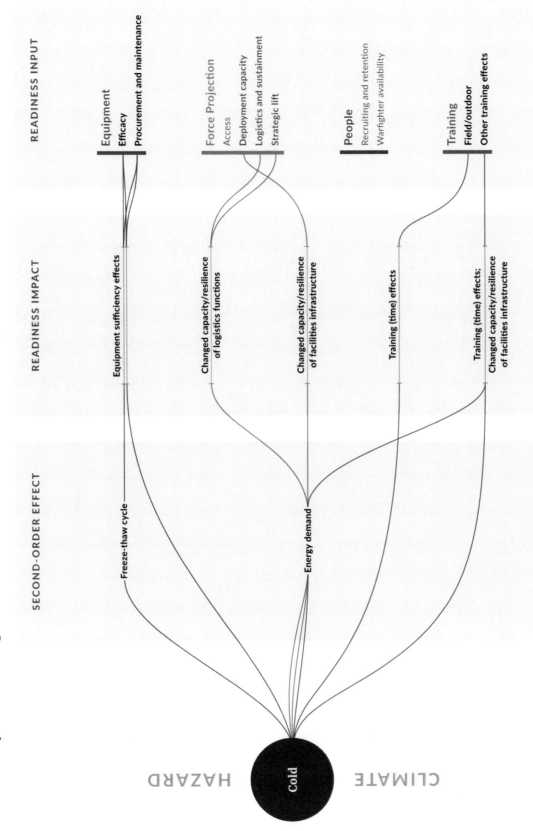

READINESS INPUT

Equipment
Efficacy
Procurement and maintenance

Force Projection
Access
Deployment capacity
Logistics and sustainment
Strategic lift

People
Recruiting and retention
Warfighter availability

Training
Field/outdoor
Other training effects

READINESS IMPACT

Equipment sufficiency effects

Changed capacity/resilience
of logistics functions

Changed capacity/resilience
of facilities infrastructure

Training (time) effects

Training (time) effects;

Changed capacity/resilience
of facilities infrastructure

SECOND-ORDER EFFECT

Freeze-thaw cycle

Energy demand

HAZARD

Cold

CLIMATE

FIGURE A.7
Ocean Anomalies, Wind, Sea Ice Retreat, and Precipitation Pathways Network Diagram

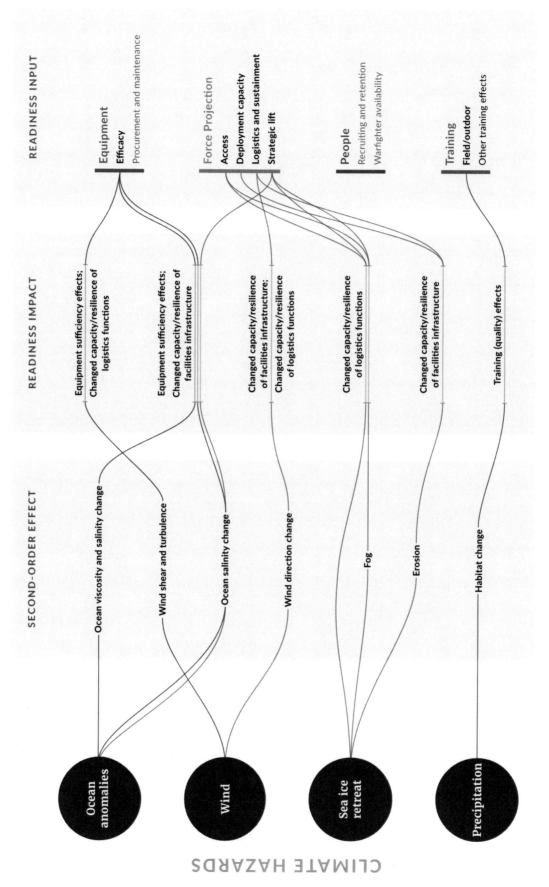

FIGURE A.8
Pathways Network Diagram for All Hazards

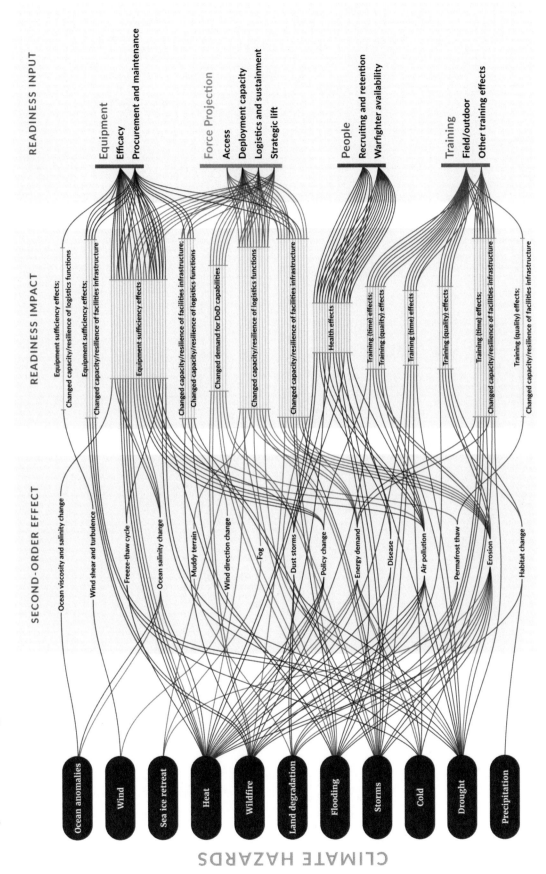

Modified Delphi Survey and Rubric

This appendix provides supplemental materials used for the modified Delphi exercise presented in Chapter 4. Table B.1 provides the rubric given to exercise participants so that they could generate their ratings. Table B.2 shows the survey sheet used to collect modified Delphi responses. Finally, Tables B.3 and B.4 show the ratings resulting from the exercise for adaptive capacity and sensitivity, respectively. Comments from the participants are paraphrased in the last column, "Details on Rating Assignment."

TABLE B.1

Rubric for Adaptive Capacity and Sensitivity Ratings

	Analysis Criteria	Questions to Consider
Hazard exposure	**High** • Multiple exposure pathways from a given hazard for a given readiness input impact exists, or • Multiple hazards with more than one pathway have a readiness input impact **Medium** • More than one exposure pathway from a given hazard for given readiness input impact exists, or • More than one hazard is associated with the readiness input impact **Low** • Singular pathway from a hazard to the readiness input impact exists, or • The hazard occurrence is limited	• Are there many hazards that create a readiness impact? • Are there multiple exposure pathways from a given hazard that create the readiness impact? (thickness of the line) • Is the exposure a permanent change/stressor or a shock? • Is the exposure confined to a specific location or does it apply to the operating environment in a region? • Exposure ratings can consider future projections of the hazard severity.
Adaptive capacity rating	**Low** • Readiness input (subcategory) has minimal or no adaptive capacity to prevent disruptions from occurring from hazard. • Limited number of options or requires extensive physical change, cost, and/or time • Example: relocating buildings outside of flood prone regions **Medium** • Readiness input (subcategory) has some adaptive capacity, but disruptions may still occur from hazard. • Limited number of options that may require some resources • Example: placing sandbags in front of buildings to mitigate flood impact **High** • Readiness input (subcategory) has adaptive capacity to minimize disruption or impacts from hazard. • Many feasible options that require limited resources • Example: schedule adjustments or rerouting assets or people **Unknown/insufficient** • Adaptive capacity for readiness input (subcategory) is unknown.	• Are there many options for adjusting? How much lead time, physical change, or costs are involved in mitigating the effects of the hazard? • Do adjustments rely on overall system capacity or redundancies and do these present minimal or significant delay?

Table B.1—Continued

	Analysis Criteria				Questions to Consider
Sensitivity rating	**Major** • Effect on readiness input (subcategory) is severe, widespread, or enduring • Nominal exposure causes major damage/safety impacts/effects that require substantial cost and time	**Moderate** • Nominal exposure causes moderate effects with potential degradation to function but can be repaired or restored with some cost or recovery time	**Minimal** • Readiness input (subcategory) experience impacts, but they are limited in scope and are transitory • Nominal exposure has minor effects with no degradation to function and that can be repaired or recovered quickly.	**Unknown/insufficient** • Readiness inputs affected but loss to readiness is not measurable	• How quickly does the readiness input (subcategory) start to feel some effects from the climate hazard? • Are these assessments location- and geography-agnostic? • Historical experience may inform your assessment. • Does nominal exposure affect function or efficacy during the period of exposure? • How significant are these effects on function or efficacy? • How quickly or how much "effort" does it take for the readiness input (category) to recover? • Consider prevalence of factors that affect sensitivity such as age or condition of buildings, equipment design standards, health and well-being of personnel, or level of active natural resource/land management.

TABLE B.2
Delphi Exercise Survey Sheet

Readiness Input (subcategory)	Readiness Impact Category	Description	Associated Hazard(s)	Hazard Exposure Rating	Adaptive Capacity Rating	Sensitivity Rating	Indicators (optional)	Qualitative Discussion
Equipment (logistics)	Changed Capacity/ Resilience of Logistics Functions	Equipment is damaged or rendered less effective due to damaged/ineffective supply chains and/or navigation routes	Wildfire; Storms (tropical); Flooding (coastal); Flooding (inland); Wind					
Equipment (sufficiency)	Equipment Sufficiency Effects	Equipment itself is damaged or rendered less effective	Heat; Cold; Drought; Land degradation; Ocean anomalies; Wildfire; Storms (tropical); Flooding (coastal); Flooding (inland); Wind					
Equipment (facilities)	Changed Capacity/ Resilience of Facilities Infrastructure	Equipment facilities are damaged	Wildfire; Drought; Storms (tropical); Flooding (coastal); Flooding (inland)					
People	Health Effects	Warfighter availability and/or ability to recruit is reduced	Heat; Drought; Land degradation; Flooding (inland); Flooding (coastal); Storms (tropical); Wildfire					
Training (natural environment)	Training (Quality) Effects	Training quality reduced due to altered natural environment	Heat; Drought; Land degradation; Precipitation (other); Flooding (coastal); Storms (tropical); Wildfire					
Training (natural environment)	Training (Time) Effects	Training time reduced due to altered natural environment	Heat; Cold; Drought; Flooding (coastal); Flooding (inland); Storms (tropical); Wildfire					
Training (facilities)	Training (Quality) Effects	Training quality reduced due to damage to facilities	Land degradation					
Training (facilities)	Training (Time) Effects	Training time reduced due to damage to facilities	Heat; Cold; Storms (winter/ ice); Storms (tropical); Land degradation; Flooding (coastal); Flooding (inland); Wildfire; Drought					

Table B.2—Continued

Readiness Input (subcategory)	Readiness Impact Category	Description	Associated Hazard(s)	Hazard Exposure Rating	Adaptive Capacity Rating	Sensitivity Rating	Indicators (optional)	Qualitative Discussion
Force Projection (facilities)	Changed Capacity/ Resilience of Facilities Infrastructure	Deployment capacity is reduced due to damage or reduced access to facilities	Heat; Cold; Flooding (coastal); Flooding (inland); Land degradation; Sea ice retreat; Wildfire; Drought; Wind					
Force Projection (energy [requirements])	Changed Capacity/ Resilience of Facilities Infrastructure	Readiness is reduced as a result of increased energy requirements of facilities	Heat; Cold					
Force Projection (infrastructure)	Changed Capacity/ Resilience of Logistics Functions	Deployment capacity is reduced due to damage to supply chain and/or transportation infrastructure	Heat; Land degradation; Storms (tropical); Flooding (coastal); Flooding (inland)					
Force Projection (natural environment)	Changed Capacity/ Resilience of Logistics Functions	Logistics capacity is reduced, disrupted, and/or more dangerous due to altered natural environment	Drought; Land degradation; Sea ice retreat; Storms (tropical); Storms (winter/ice); Wind					
Force Projection (energy & water [requirements])	Changed Capacity/ Resilience of Logistics Functions	Readiness is reduced as a result of increased energy and/or water requirements of deployed forces	Heat; Cold; Drought					
Force Projection (equipment)	Equipment Sufficiency Effects	Strategic lift capacity is reduced as a result of damaged or ineffective equipment	Heat; Ocean anomalies; Wildfire					
Force Projection (demand)	Changed Demand for DoD Capabilities	Deployment capacity is reduced due to increased demand for HADR missions	Storms (tropical); Wildfire					

TABLE B.3

Adaptive Capacity Assessment, with Comments from Participants

Readiness Input and Impact	Readiness Input and Description	Rating	Details on Rating Assignment
Training (time) effects	(Training: Natural environment) Training time is reduced due to altered natural environment	Low to medium to high (Low, Low, Medium, Medium, Medium, High)	• Low: Options for preventing damage or degradation may be limited for most hazards, as natural environments are not easily recoverable or may take extensive time (years) to recover. • Medium: Services may be able to reschedule or relocate, but that may take away from training time. • High: Training activities may be able to take place in other locations. Monitoring of the natural environment for changes or potential degradation may help with adaptation.
Training (quality) effects	(Training: Facilities) Training quality is reduced due to damage to facilities	Low to medium to high (Low, Low, Medium, Medium, Medium, High)	• Low: Moving the facilities or infrastructure may be very costly or training quality may be adversely affected because of the unique nature of one-of-a-kind–type facilities that may not be readily recoverable or substitutable. • Medium: Some training can be relocated to other facilities. • High: In most cases, multiple facilities can provide a needed training environment, so temporary or even permanent shifts in training location are possible.
Training (time) effects	(Training: Facilities) Training time is reduced due to damage to facilities	Low to medium to high (Low, Low, Medium, Medium, High, High)	• Low: Some specialized training may be location-specific, and damage to training facilities may affect training time without an ability to recover in a timely fashion. • Medium: The impact to training time may depend on the breadth and severity of the event. • High: Training activities may be able to take place elsewhere, thereby reducing the criticality of training at any one location.
Training (quality) effects	(Training: Natural environment) Training quality is reduced due to altered natural environment	Low to medium (Low, Low, Medium, Medium, Medium, Medium)	• Low: It may be difficult or impossible to replace the ability to train in unique environments that are permanently altered. • Medium: It might not be possible to relocate training, but training could be rescheduled or could be supplemented or replaced by simulated training. Limited options that include active management or exposure prevention/protection.
Changed capacity and resilience of facilities infrastructure	(Equipment: Facilities) Equipment facilities are damaged	Low to medium (Low, Low, Medium, Medium, Medium, Medium)	• Low: Equipment can be moved, but there might eventually be a lack of places to move equipment. • Medium: Some adaptation measures are less expensive (e.g., sandbags, fire lines), while others are expensive or cost-prohibitive (e.g., seawalls).
Changed capacity and resilience of facilities infrastructure	(Force projection: Facilities) Deployment capacity is reduced due to damage or reduced access to facilities	Low to medium (Low, Low, Medium, Medium, Medium, Medium)	• Low: Depending on the criticality of timing needed for force projection, operational impacts due to untimely facility losses may be extreme and even prevent expedient power projection. • Medium: Some adaptation options exist to mitigate the most-severe consequences (e.g., seawalls, burn lines); location and design standards are fixed in time, but some "hardening" measures are readily available.

Table B.3—Continued

Readiness Input and Impact	Readiness Input and Description	Rating								Details on Rating Assignment
Equipment sufficiency effects	(Equipment: Sufficiency) Equipment itself is damaged or rendered less effective	Low to medium	Low	Low	Low	Medium	Medium	Medium	Medium	• Low: There are limited options available for such equipment as sensors and communication systems; new acquisition would be required. • Medium: Numerous adaptation options exist, but, if exposure reaches a "new normal" at the design threshold, whole classes of equipment may no longer be effective and may have to be completely replaced.
Equipment sufficiency effects	(Force projection: Equipment) Strategic lift capacity is reduced as a result of damaged or ineffective equipment	Medium	Medium	Medium	Medium	Medium	Medium	Medium	Medium	• Medium: Strategic lift capacity may be more adaptable because different routes or means can be considered, but there is still the issue of a long equipment time horizon. Increased maintenance and scheduling changes are options for adaptation.
Changed capacity and resilience of logistics functions	(Force projection: Energy and water requirements) Readiness is reduced as a result of increased energy and/or water requirements of deployed forces	Medium to high	Unknown	Medium	Medium	Medium	Medium	Medium	High	• Medium: Reduction of water or energy supply can be resolved through increases in redundancies, resilience, and backup supplies for forces. • High: There are likely several options for providing water (fewer for energy). But overall, there are technology and capacity-building adaptive options, as well as commercial variants to supplement military capabilities.
Changed capacity and resilience of facilities infrastructure	(Force projection: Energy requirements) Readiness is reduced as a result of increased energy requirements of facilities	Medium to high	Medium	Medium	Medium	Medium	Medium	High	High	• Medium: Options exist to increase energy supply, but implementation may be challenged by grid capacity or technical constraints. • High: It might be possible to shift high energy usage in time for most events or prioritize energy usage to focus on critical activities.
Changed capacity and resilience of logistics functions	(Force projection: Natural environment) Logistics capacity is reduced, disrupted, and/or more dangerous due to altered natural environment	Medium to high	Medium	Medium	Medium	High	High	High	High	• Medium: Impacts to logistics capacity may depend on the breadth and severity of the event. For some areas, it may be easy to plan logistics to avoid exposure; for other areas, it may be difficult or impossible. • High: There are likely many options for rerouting; there are multiple means of access that can be considered.

Table B.3—Continued

Readiness Input Impact	Readiness Input and Description	Rating	Details on Rating Assignment
Changed demand for DoD capabilities	(Force projection: Demand) Deployment capacity is reduced due to increased demand for HADR missions	Medium to high; Medium Medium Medium High High High High	• Medium: There is significant capacity across active and reserve forces, but demands could be high. • High: Assistance is a matter of policy, suggesting that there is potential for high adaptability and low sensitivity.
Changed capacity and resilience of logistics functions	(Force projection: Infrastructure) Deployment capacity is reduced due to damage to supply chain and/or transportation infrastructure	Medium to high; Medium Medium Medium High High High High	• Medium: The impact to deployment capacity may depend on the breadth and severity of the event. • High: There are alternative transport means for certain items; assuming events are not concurrent or do not affect a critical node, rescheduling could occur.
Health effects	(People) Warfighter availability and/or ability to recruit is reduced	Medium to high; Medium Medium Medium High High High High	• Medium: Monitoring health during extreme heat events and using cooling methods (e.g., using cooling centers, increasing water-supply reserve) can help warfighters with extreme heat and drought hazards. • High: Many options exist, including relocating, rescheduling, and providing protective measures.
Changed capacity and resilience of logistics functions	(Equipment: Logistics) Equipment is damaged or rendered less effective due to damaged or ineffective supply chains and/or navigation routes	Medium to high; Medium Medium High High High High High	• Medium: Some supply chains and navigation routes might be critical, while others can be avoided or replaced. • High: Disruptions could be minor because services could choose whether to use the equipment while looking for alternative supply chains.

TABLE B.4

Sensitivity Assessment with Comments

Readiness Input Impact	Readiness Input and Description	Rating		Details on Rating Assignment
Changed capacity and resilience of logistics functions	(Equipment: Logistics) Equipment is damaged or rendered less effective due to damaged or ineffective supply chains and/or navigation routes	Minimal to moderate	Moderate, Moderate, Moderate, Moderate, Moderate, Minimal, Minimal	• Minimal: Hazards listed are distinct events that cause temporary disruptions. It is unlikely that the hazards will affect a majority of the routes at the same time. • Moderate: It depends on the hazard and the location, but there will be impacts if the affected route or supply chain is critical.
Changed capacity and resilience of logistics functions	(Force projection: Energy and water requirements) Readiness is reduced as a result of increased energy and/or water requirements of deployed forces	Minimal to moderate	Moderate, Moderate, Moderate, Minimal, Minimal, Minimal, Unknown	• Minimal: There is an assumption that energy and water requirements would differ greatly from today's requirements, and the systems should have enough slack capacity to respond. Requirements could, however, be an issue in forward operating environment with frequent maneuvers. • Moderate: Some natural environments will change rapidly because of climate change, while others will change more slowly.
Changed demand for DoD capabilities	(Force projection: Demand) Deployment capacity is reduced due to increased demand for HADR missions	Minimal to moderate	Moderate, Minimal, Minimal, Minimal, Minimal, Minimal, Minimal	• Minimal: The consensus is that DSCA and HADR often do not require large amounts of resources, responsibility is shared with aid organizations and other partners, and the resources required may not significantly reduce the capacity available for other missions. • Moderate: A shift in the overall mission set to include increased HADR missions and requirements for responsiveness for such missions will have far-reaching effects on readiness cycles, stress on the force, and potential capability and acquisition decisions, diverting resources to these missions even if the actualized demand is small.
Changed capacity and resilience of logistics functions	(Force projection: Natural environment) Logistics capacity is reduced, disrupted, and/or more dangerous due to altered natural environment	Moderate	Moderate, Moderate, Moderate, Moderate, Moderate, Moderate, Moderate	• Moderate: The consensus is for moderate, but, to some extent, this depends on how critical the timing of operations is.
Health effects	(People) Warfighter availability and/or ability to recruit is reduced	Moderate to major	Major, Moderate, Moderate, Moderate, Moderate, Moderate, Moderate	• Moderate: Health effects may be catastrophic from some hazards and minimal from others. • Major: Without adaptation, even low-grade exposures will have health effects. Catastrophic hazards, such as wildfires and floods, are life-threatening.

Table B.4—Continued

Readiness Input and Impact	Readiness Input and Description		Rating				Details on Rating Assignment
Training (time) effects	(Training: Facilities) Training time is reduced due to damage to facilities	Moderate to major	Moderate	Moderate	Moderate	Major	Major	• Moderate: Built structures are often older and not well maintained, thereby increasing sensitivity. Most installations experience damage that can be repaired, although some have experienced severe damage, and, in the future, this could be worse as weather diverges from historical experience on which design standards are based. • Major: A high degree of land degradation can eliminate a training facility.
Equipment sufficiency effects	(Force projection: Equipment) Strategic lift capacity is reduced as a result of damaged or ineffective equipment	Moderate to major	Moderate	Moderate	Moderate	Major	Major	• Moderate: Effects depend on the platform, but performance design requirements are fixed. • Major: Impacts, such as alternative fuel mandates, overheating aircraft, and lowered visibility from smoke, may be felt immediately.
Changed capacity and resilience of logistics functions	(Force projection: Infrastructure) Deployment capacity is reduced due to damage to supply chain and/ or transportation infrastructure	Moderate to major	Unknown	Moderate	Moderate	Major	Major	• Moderate: Logistics functions may have built-in redundancy (e.g., alternate routes). • Major: If enough disruptions occur, redundancy may become insufficient. Adaptive capacity may be limited, since needed repairs could involve long timelines and long investments or be out of DoD's control.
Training (quality) effects	(Training: Natural environment) Training quality is reduced due to altered natural environment	Moderate to major	Moderate	Moderate	Major	Major	Major	• Moderate: Effects depend on the breadth and severity of the event, so "moderate" was used as an average. Ecosystems are resilient to hazards to some degree, assuming that there is some active management of natural resources. • Major: In some cases, the natural environment is critical to training. If hazards are severe and far-reaching or affect a unique range, then sensitivity can be high.
Training (time) effects	(Training: Natural environment) Training time is reduced due to altered natural environment	Moderate to major	Moderate	Moderate	Major	Major	Major	• Moderate: The assumed scenario is that either the hazard has limited impact on the range or the range's natural-resource condition is not an essential element of the training. • Major: The assumed scenario is that the hazard's impact on the natural environment is strong or the natural environment's condition is an essential element of training.
Training (quality) effects	(Training: Facilities) Training quality is reduced due to damage to facilities	Moderate to major	Moderate	Moderate	Major	Major	Major	• Moderate: Active investments in Sustainable Range Program activities could reduce sensitivity (within limits). • Major: A high degree of land degradation can eliminate a training facility.

Table B.4—Continued

Readiness Input Impact	Readiness Input and Description	Rating								Details on Rating Assignment
Equipment sufficiency effects	(Equipment: Sufficiency) Equipment itself is damaged or rendered less effective	Moderate to major	Moderate	Moderate	Moderate	Major	Major	Major	Major	• Moderate: Effects depend on equipment type and design thresholds. • Major: Equipment performance requirements may be robust, but they are fixed in time, and it is not clear whether climate and weather projections are included. There is experience with dust storms and extreme heat delaying operations and damaging equipment. Equipment failure is major and affects multiple services.
Changed capacity and resilience of facilities infrastructure	(Equipment: Facilities) Equipment facilities are damaged	Moderate to major	Moderate	Moderate	Moderate	Major	Major	Major	Major	• Moderate: Built structures are often older and not well maintained, increasing sensitivity. But some buildings that experience damage can be repaired, although some have experienced severe damage, and, in the future, this could be worse as weather diverges from historical experience on which design standards are based. • Major: Historical experience with hazards illustrates that facilities experience physical damage and degradation.
Changed capacity and resilience of facilities infrastructure	(Force projection: Facilities) Deployment capacity is reduced due to damage or reduced access to facilities	Moderate to major	Moderate	Moderate	Moderate	Major	Major	Major	Major	• Moderate: Damage can be repaired, but future damage could be greater as weather diverges from historical experience on which design standards are based. • Major: The hazards are relatively infrequent but have potentially catastrophic consequences that may require total rebuilds.
Changed capacity and resilience of facilities infrastructure	(Force projection: Energy requirements) Readiness is reduced as a result of increased energy requirements of facilities	Minimal to moderate to major	Minimal	Minimal	Moderate	Moderate	Moderate	Major	Major	• Minimal: Effects depend on assumptions regarding overall energy grid capacity and opportunities to leverage the commercial sector. • Moderate: Effects are immediate but temporary. • Major: Because energy is required for all installations and services, changes in requirements can have substantial effects across the board.

Additional RDIM Pathway Annotations

This appendix provides further examples of annotation of RDIM-related products informed by the framework and pathways presented in this report. Specifically, this appendix shows additional annotations of the R-rating aviation pipeline model (i.e., proof-of-concept RDIM), as well as a second potential aviation pipeline with the logic diagrammed out for personnel readiness (i.e., P-rating).

Figure C.1 shows the seven climate narratives (comprising 14 climate hazard pathways) that were identified as relevant to opening work orders, meaning they would lead to something that would require an aircraft to need maintenance or repair. These seven narratives cover a variety of hazards, all of which affect the equipment readiness input (see the blue arrows). These hazards could cause damage to an aircraft while it is in flight or collateral damage on an installation. The narratives are also characterized as stressor- or shock-type hazards (see the yellow and red arrows). The hazards are predominantly shocks, although there are some hazards caused by cold and heat that could slowly cause damage.

In Figure C.2, we highlight the potential data and indicators for the seven climate narratives relevant to the opening of work orders. Similar to the flight hours example in Chapter 5, the WXX (weather) devia-

FIGURE C.1
Climate Hazard Narratives Relevant to Opening Work Orders

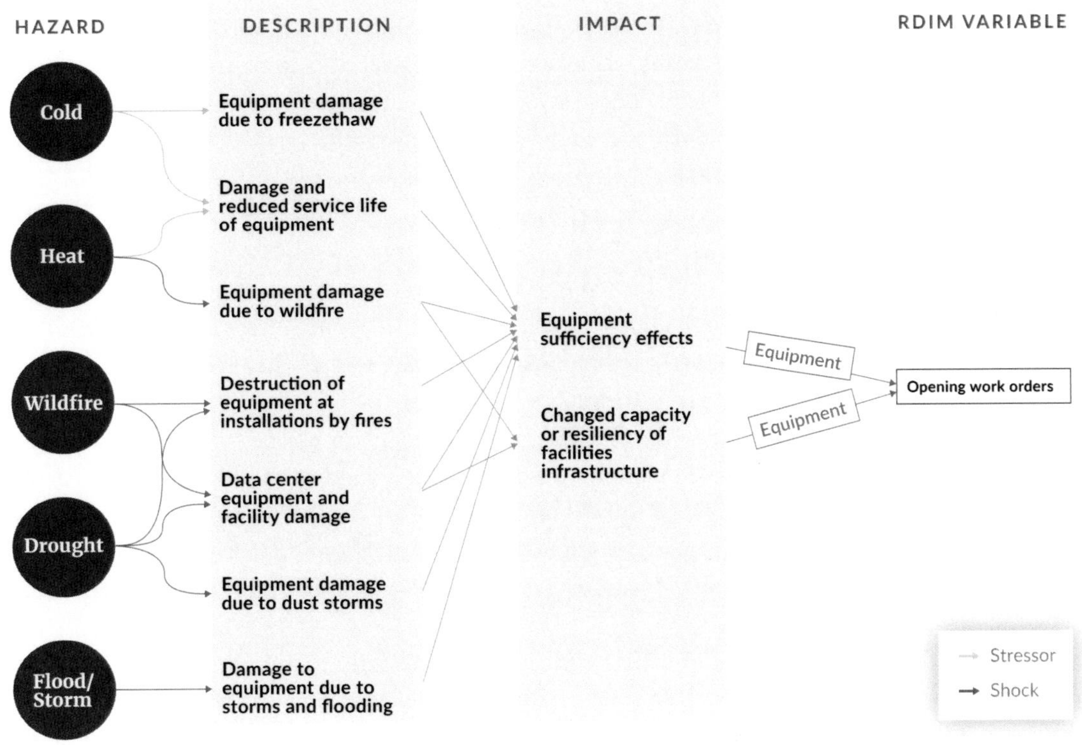

FIGURE C.2

Data and Indicators for Narratives Relevant to Opening Work Orders

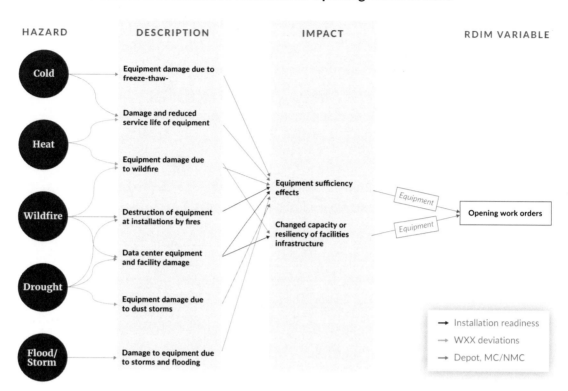

tions (green arrows) could potentially indicate that a weather event caused damage and was the reason for the deviation in the flight plan. Most of the other indicators are related to depot or to the aircraft's MC or NMC status (see the pink arrows), which is logical because these are common maintenance-related metrics tracked by aviation communities in DoD. Weather deviations, depot, and MC or NMC status are all tracked in LIMS-EV for the U.S. Air Force. The other data (see the purple arrows) come from installation-readiness sources, which were discussed in Chapter 5. These would track the relevant pathways that involve damage to and at an installation or facility. The source for these types of data would be DRRS-S reports from installations or Installation Health Assessments.

The final annotation to the R-rating production pipeline is that of maintenance throughput, which can be thought of as the efficiency rate at which maintenance work orders are completed and closed. In Figure C.3, we show the two identified climate narratives (comprising seven climate hazard pathways)—*destruction of key locations in supply chains* and *delayed maintenance of equipment*—which were caused by wildfire, floods and storms, and present a mix of shock and stressor hazards. Relative to the other annotations, there are fewer narratives that directly affect maintenance throughput; however, we note that most of the maintenance-related impacts will be those that increase the demand for maintenance (i.e., opening work orders). The identified narratives relevant to maintenance throughput would create maintenance delays and thus decrease the maintenance throughput. In Figure C.4, we show that the indicators for the climate risks related to maintenance throughput are all relevant to the pathway's effects (i.e., not directly tied to the hazard itself) and speak to the maintenance status or rates, such as MC rates and the different NMC rates, and installation readiness when facilities are involved.

Lastly, in addition to the potential aviation readiness-production pipeline for the T-rating, described in Chapter 5, we show another potential aviation readiness-production pipeline for the P-rating that could be used both in RDIM generally and to facilitate additional climate hazard pathway annotation in future

FIGURE C.3

Climate Hazard Narratives Relevant to Maintenance Throughput

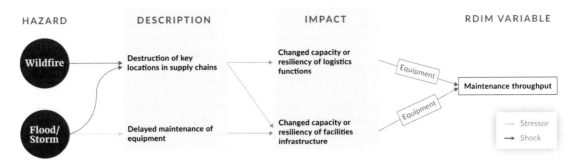

FIGURE C.4

Data and Indicators for Narratives Relevant to Maintenance Throughput

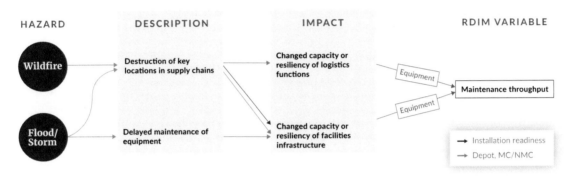

analysis. This pipeline is relatively simple and has key variables that are different from those of both the R-rating and proposed T-rating pipelines (see Figure C.5). Starting from the left in the figure, there are the *decisions*, or policy levers, available for RDIM scenario building. For personnel readiness, we included changes to authorized personnel and assigned personnel, the former relating to defined mission requirements and the latter resulting from recruitment and assignment decisions. The additional *impact* variable, aside from *authorized personnel* and *assigned personnel*, is the *crew ratio*, which represents the number of aircrews authorized per aircraft and dictates the availability of personnel to conduct operations and training. A larger crew ratio allows more time for crews to complete training between operations. The training aspect (i.e., *Ready Aircrew Program training completion*) is included because training correlates to experience levels in a unit, and the P-rating of a unit is not simply a matter of having the right number of people: It also requires having the right number of personnel with certain experience levels.[1] The step between crew ratio and Ready Aircrew Program training completion is the only one where some sophisticated accounting or even modeling may be necessary in RDIM development (note the blue arrow in the figure); the other steps are simple calculations based primarily on preestablished business rules. The final step, where personnel availability impact on P-rating is calculated, would compare the number of available and training personnel with the number of authorized personnel. This is a simplified view of the personnel readiness-production pipeline and does not fully capture more of the "upstream" activities, such as recruiting and retention. If additional variables are

---

[1] The Ready Aircrew Program is an example set of training requirements used by some parts of the aviation community in the U.S. Air Force.

FIGURE C.5

Potential Aviation Personnel Readiness-Production Pipeline

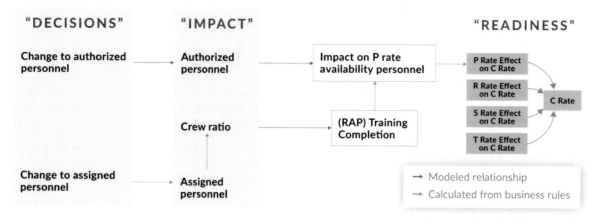

SOURCE: Features information from Deputy Assistant Secretary of Defense for Force Readiness and RDIM Modeling Team, email communications and discussions with the authors, February 3 and March 19, 2022.
NOTE: RAP = Ready Aircrew Program.

needed for personnel-related readiness scenarios in RDIM, there are previous studies in this area to inform the choice of variables.[2]

---

[2] Examples of relevant work include Carl J. Dahlman, Robert Kerchner, and David E. Thaler, *Setting Requirements for Maintenance Manpower in the U.S. Air Force*, RAND Corporation, MR-1436-AF, 2002; Albert A. Robbert, Anthony D. Rosello, Clarence R. Anderegg, John A. Ausink, James H. Bigelow, William W. Taylor, and James Pita, *Reducing Air Force Fighter Pilot Shortages*, RAND Corporation, RR-1113-AF, 2015; Albert A. Robbert, Patricia K. Tong, and Chaitra M. Hardison, *Retention of Enlisted Maintenance, Logistics, and Munitions Personnel: Analysis and Results*, RAND Corporation, RR-A546-1, 2022.

Abbreviations

AHTA	All-Hazard Threat Assessment
ARRM	Army Range Requirements Model
ASRA	Army Strategic Readiness Assessment
CCI	Climate Change Impacts
CONUS	continental United States
DAF	Department of the Air Force
DCAT	Department of Defense Climate Assessment Tool
DoD	Department of Defense
DRRS	Defense Readiness Reporting System
DRRS-A	Defense Readiness Reporting System–Army
DRRS-S	Defense Readiness Reporting System–Strategic
DSCA	Defense Support of Civil Authorities
E&ER	Environment and Energy Resilience
FY	fiscal year
HADR	humanitarian assistance and disaster relief
IPCC	Intergovernmental Panel on Climate Change
ISR	Installation Status Report
JCCA	Joint Combat Capabilities Assessment
LIMS-EV	Logistics, Installations and Mission Support-Enterprise View
MA	mission assurance
MC	mission-capable
NMC	non–mission-capable
OSD	Office of the Secretary of Defense
P&R	Personnel and Readiness
RCP	Representative Concentration Pathway
RDIM	Readiness Decision Impact Model
SDDC	Surface Deployment and Distribution Command
USINDOPACOM	U.S. Indo-Pacific Command
WOWA	Weighted Ordered Weighted Average

References

American Security Project, homepage, undated. As of August 5, 2022:
https://www.americansecurityproject.org/

Austin, Lloyd J., III, "Statement by Secretary of Defense Lloyd J. Austin III on Tackling the Climate Crisis at Home and Abroad," press release, U.S. Department of Defense, January 27, 2021.

Bela, Joe, "AFIMSC Analytics Tool Expands, Helps Leaders Prioritize Investments," press release, May 28, 2020.

Best, Katharina Ley, Scott R. Stephenson, Susan A. Resetar, Paul W. Mayberry, Emmi Yonekura, Rahim Ali, Joshua Klimas, Stephanie Stewart, Jessica Arana, Inez Khan, and Vanessa Wolf, *Climate and Readiness: Understanding Climate Vulnerability of U.S. Joint Force Readiness: Climate Hazard Pathways Appendix*, RAND Corporation, RR-A1551-2, 2023. As of June 1, 2023:
https://www.rand.org/pubs/research_reports/RRA1551-1.html

Biden, Joseph R., Jr., "Tackling the Climate Crisis at Home and Abroad," Executive Order 14008, Executive Office of the President, February 1, 2021a.

Biden, Joseph R., Jr., *Interim National Security Strategic Guidance*, The White House, March 2021b.

Bye, Hilde-Gunn, "The USA Will Release a New National Arctic Strategy," *High North News*, last updated April 8, 2022.

Center for Climate and Security, homepage, undated. As of August 5, 2022:
https://climateandsecurity.org/

Center for Excellence in Disaster Management and Humanitarian Assistance, "Climate Change Impacts," webpage, undated. As of July 1, 2022:
https://www.cfe-dmha.org/Programs/Climate-Change-Impacts

Chairman of the Joint Chiefs of Staff, *CJCS Guide to the Chairman's Readiness System*, CJCS Guide 3401D, November 15, 2010.

CNA Corporation, *National Security and the Threat of Climate Change*, 2007.

Council on Strategic Risks and Center for Climate and Security, *Climate Change and the National Defense Authorization Act*, June 2022.

DAF—*See* Department of the Air Force.

Dahlman, Carl J., Robert Kerchner, and David E. Thaler, *Setting Requirements for Maintenance Manpower in the U.S. Air Force*, RAND Corporation, MR-1436-AF, 2002. As of July 13, 2022:
https://www.rand.org/pubs/monograph_reports/MR1436.html

Defense Science Board, *Report of the Defense Science Board Task Force on Trends and Implications of Climate Change for National and International Security*, Office of the Under Secretary of Defense for Acquisition, Technology, and Logistics, U.S. Department of Defense, October 2011.

Department of the Air Force, *Mission Assurance*, Air Force Policy Directive 10-24, November 5, 2019.

Department of the Air Force, *The Department of the Air Force Arctic Strategy*, July 21, 2020a.

Department of the Air Force, *Force Readiness Reporting*, Air Force Instruction 10-201, December 22, 2020b.

Department of the Army, *The Army Sustainable Range Program*, Army Regulation 350-19, August 30, 2005.

Department of the Army, *Defense Readiness Reporting System—Army Procedures*, Pamphlet 220-1, November 16, 2011.

Department of the Army, *The Army Protection Program*, Army Regulation 525-2, December 8, 2014.

Department of the Army, *FY18 ISR Infrastructure Implementing Instructions*, Version 1.0, October 1, 2017.

Department of the Army, *Installation Status Report Program*, Army Regulation 210-14, June 11, 2019.

Department of the Army, *Army Strategic and Operational Readiness*, Army Regulation 525-30, April 9, 2020.

Department of the Army, *Regaining Arctic Dominance*, Chief of Staff Paper No. 3, January 19, 2021.

Department of the Army, Office of the Assistant Secretary of the Army for Installations, Energy and Environment, *United States Army Climate Strategy*, February 2022.

Department of the Navy, *Defense Readiness Reporting System–Navy (DRRS-N) Overview Course–PESTO*, CVN DRRS training facilitator's guide, undated.

Department of the Navy, *Defense Readiness Reporting System–Navy*, OPNAV Instruction 2502.360A, October 17, 2014.

Department of the Navy, *Navy Mission Assurance Program*, OPNAV Instruction 3502.8, November 8, 2017.

Department of the Navy, *A Blue Arctic: A Strategic Blueprint for the Arctic*, January 5, 2021.

Department of the Navy, Headquarters U.S. Marine Corps, *Marine Corps Mission Assurance*, Marine Corps Order 3058.1, October 23, 2014.

Department of the Navy, Office of the Assistant Secretary of the Navy for Energy, Installations, and Environment, *Department of the Navy Climate Action 2030*, May 2022.

Federal Emergency Management Agency, National Risk Index, "Coastal Flooding," webpage, undated-a. As of April 26, 2023:
https://hazards.fema.gov/nri/coastal-flooding

Federal Emergency Management Agency, National Risk Index, "Drought," webpage, undated-b. As of April 26, 2023:
https://hazards.fema.gov/nri/drought

Federal Emergency Management Agency, National Risk Index, "Wildfire," webpage, undated-c. As of April 26, 2023:
https://hazards.fema.gov/nri/wildfire

Gold, Dara, Bart E. Bennett, Bradley DeBlois, Ronald G. McGarvey, and Anna Jean Wirth, *A Modeling Framework for Optimizing U.S. Air Force Fighter Pilot Access to Advanced Training Ranges*, RAND Corporation, TL-A169-1, 2020. As of July 13, 2022:
https://www.rand.org/pubs/tools/TLA169-1.html

Health.mil, "Medical Surveillance Monthly Report," webpage, last updated 2022. As of April 15, 2022:
https://health.mil/Military-Health-Topics/Health-Readiness/AFHSD/Reports-and-Publications/Medical-Surveillance-Monthly-Report

Herrera, G. James, *The Fundamentals of Military Readiness*, Congressional Research Service, R46559, October 2, 2020.

Intergovernmental Panel on Climate Change, "Summary for Policymakers," in Valérie Masson-Delmotte, Panmao Zhai, Anna Pirani, Sarah L. Connors, C. Péan, Sophie Berger, N. Caud, Y. Chen, Leah Goldfarb, Melissa I. Gomis, et al., eds., *Climate Change 2021: The Physical Science Basis: Contribution of Working Group I to the Sixth Assessment Report of the Intergovernmental Panel on Climate Change*, Cambridge University Press, 2021.

Intergovernmental Panel on Climate Change, "Summary for Policymakers," in Hans-Otto Pörtner, Debra C. Roberts, Melinda M. B. Tignor, Elvira Poloczanska, Katja Mintenbeck, Andrés Alegría, Marlies Craig, Stefanie Langsdorf, Sina Löschke, Vincent Möller, et al., eds., *Climate Change 2022: Impacts, Adaptation and Vulnerability: Working Group II Contribution to the Sixth Assessment Report of the Intergovernmental Panel on Climate Change*, Cambridge University Press, 2022.

IPCC—*See* Intergovernmental Panel on Climate Change.

Kaufman, Holly, and Sherri Goodman, *Climate Change in the U.S. National Security Strategy: History and Recommendations*, Council on Strategic Risks and Center for Climate and Security, Briefer No. 1, June 29, 2021.

Lachman, Beth E., Susan A. Resetar, Nidhi Kalra, Agnes Gereben Schaefer, and Aimee E. Curtright, *Water Management, Partnerships, Rights, and Market Trends: An Overview for Army Installation Managers*, RAND Corporation, RR-933-A, 2016. As of January 10, 2022:
https://www.rand.org/pubs/research_reports/RR933.html

Lachman, Beth E., Anny Wong, and Susan A. Resetar, *The Thin Green Line: An Assessment of DoD's Readiness and Environmental Protection Initiative to Buffer Installation Encroachment*, RAND Corporation, MG-612-OSD, 2007. As of January 10, 2022:
https://www.rand.org/pubs/monographs/MG612.html

Leidos, Inc., *Climate Change Planning Handbook: Installation Adaptation and Resilience*, prepared for Naval Facilities Engineering Command, January 2017.

Lempert, Robert, Jeffrey Arnold, Roger Pulwarty, Kate Gordon, Katherine Greig, Cat Hawkins Hoffman, Dale Sands, and Caitlin Werrell, "Reducing Risks Through Adaptation Actions," in David Reidmiller, Christopher W. Avery, David R. Easterling, Kenneth E. Kunkel, Kristin Lewis, Thomas K. Maycock, and Brooke C. Stewart, eds., *Fourth National Climate Assessment:* Vol. II, *Impacts, Risks, and Adaptation in the United States*, U.S. Global Change Research Program, 2018.

Myers, Meghann, "New Data Model Predicts How Deployments Affect Future Readiness," *Air Force Times*, December 29, 2022.

Narayanan, Anu, Michael J. Lostumbo, Kristin Van Abel, Michael T. Wilson, Anna Jean Wirth, and Rahim Ali, *Grounded: An Enterprise-Wide Look at Department of the Air Force Installation Exposure to Natural Hazards: Implications for Infrastructure Investment Decisionmaking and Continuity of Operations Planning*, RAND Corporation, RR-A523-1, 2021. As of August 5, 2022:
https://www.rand.org/pubs/research_reports/RRA523-1.html

National Intelligence Council, *National Intelligence Estimate: Climate Change and International Responses Increasing Challenges to US National Security Through 2040*, NIC-NIE-2021-10030-A, October 2021.

National Weather Service, "Flood and Flash Flood Definitions," webpage, undated. As of April 26, 2023:
https://www.weather.gov/mrx/flood_and_flash

National Weather Service, *Storm Data Preparation*, National Weather Service Instruction 10-1605, July 26, 2021.

Naval Facilities Engineering Command, *Climate Change Installation Adaptation and Resilience*, January 2017.

Obama, Barack, "Federal Leadership in Environmental, Energy, and Economic Performance," Executive Order 13514, Executive Office of the President, October 5, 2009.

Office of Inspector General, *Evaluation of the Department of Defense's Efforts to Address the Climate Resilience of U.S. Military Installations in the Arctic and Sub-Arctic*, U.S. Department of Defense, DODIG-2022-083, April 13, 2022.

Pinson, A. O., K. D. White, S. A. Moore, S. D. Samuelson, B. A. Thames, P. S. O'Brien, C. A. Hiemstra, P. M. Loechl, and E. E. Ritchie, *Army Climate Resilience Handbook*, U.S. Army Corps of Engineers, 2020.

Pinson, A. O., K. D. White, E. E. Ritchie, H. M. Conners, and J. R. Arnold, *DOD Installation Exposure to Climate Change at Home and Abroad*, U.S. Army Corps of Engineers, April 2021.

Public Law 116-283, William M. (Mac) Thornberry National Defense Authorization Act for Fiscal Year 2021, January 1, 2021.

Readiness and Environmental Protection Integration, homepage, U.S. Department of Defense, undated. As of April 15, 2022:
https://www.repi.mil/

Reisinger, Andy, Mathias Garschagen, Katharine J. Mach, Minal Pathak, Elvira Poloczanska, Maarten van Aalst, Alexander C. Ruane, Mark Howden, Margot Hurlbert, Katja Mintenbeck, et al., *The Concept of Risk in the IPCC Sixth Assessment Report: A Summary of Cross-Working Group Discussions*, Intergovernmental Panel on Climate Change, September 4, 2020.

Resetar, Susan A., and Neil Berg, *An Initial Look at DoD's Activities Toward Climate Change Resiliency: An Annotated Bibliography*, RAND Corporation, WR-1140-AF, 2016. As of August 11, 2022:
https://www.rand.org/pubs/working_papers/WR1140.html

Robbert, Albert A., Anthony D. Rosello, Clarence R. Anderegg, John A. Ausink, James H. Bigelow, William W. Taylor, and James Pita, *Reducing Air Force Fighter Pilot Shortages*, RAND Corporation, RR-1113-AF, 2015. As of July 13, 2022:
https://www.rand.org/pubs/research_reports/RR1113.html

Robbert, Albert A., Patricia K. Tong, and Chaitra M. Hardison, *Retention of Enlisted Maintenance, Logistics, and Munitions Personnel: Analysis and Results*, RAND Corporation, RR-A546-1, 2022. As of July 13, 2022: https://www.rand.org/pubs/research_reports/RRA546-1.html

Sacha, Dominik, Hansi Senaratne, Bum Chol Kwon, Geoffrey Ellis, and Daniel A. Keim, "The Role of Uncertainty, Awareness, and Trust in Visual Analytics," *IEEE Transactions on Visualization and Computer Graphics*, Vol. 22, No. 1, 2016.

SDDC—*See* Surface Deployment and Distribution Command.

Secretary of Defense, "Establishment of the Climate Working Group," memorandum for senior Pentagon leadership, commanders of the combatant commands, and defense agency and DoD field activity directors, March 9, 2021.

Secretary of the Army, "Army Directive 2020-08 (U.S. Army Installation Policy to Address Threats Caused by Changing Climate and Extreme Weather)," memorandum for U.S. Army North, U.S. Army South, U.S. Army Africa/Southern European Task Force, et al., September 11, 2020.

Shaw, Frederick J., ed., *Locating Air Force Base Sites: History's Legacy*, Air Force History and Museums Program, 2004.

Surface Deployment and Distribution Command Transportation Engineering Agency, *Port Look 2021*, May 10, 2021, Not available to the general public.

U.S. Army War College, *How the Army Runs: A Senior Leader Reference Handbook*, January 29, 2020.

U.S. Climate Resilience Toolkit, "Steps to Resilience Overview," webpage, last updated July 28, 2022. As of August 4, 2022: https://toolkit.climate.gov/steps-to-resilience/steps-resilience-overview

U.S. Code, Title 10, Armed Forces; Subtitle A, General Military Law; Part I, Organization and General Military Powers; Chapter 5, Joint Chiefs of Staff; Section 153, Chairman: Functions.

U.S. Department of Defense, *Strategic Sustainability Performance Plan for Fiscal Year 2010*, August 26, 2010.

U.S. Department of Defense, *Mission Assurance Strategy*, April 2012.

U.S. Department of Defense, *2014 Climate Change Adaptation Roadmap*, Office of the Assistant Secretary of Defense (Energy, Installations & Environment), June 2014.

U.S. Department of Defense, *Climate Change Adaptation and Resilience*, DoD Directive 4715.21, January 14, 2016.

U.S. Department of Defense, *2017 Sustainable Ranges*, Under Secretary of Defense (Personnel and Readiness), May 2017.

U.S. Department of Defense, *Department of Defense Readiness Reporting System (DRRS)*, DoD Directive 7730.65, May 31, 2018a.

U.S. Department of Defense, *Mission Assurance (MA)*, DoD Directive 3020.40, September 11, 2018b.

U.S. Department of Defense, *Department of Defense Arctic Strategy*, Office of the Under Secretary of Defense for Policy, June 2019.

U.S. Department of Defense, "DoD Climate Assessment Tool," fact sheet, April 5, 2021a.

U.S. Department of Defense, *Department of Defense Climate Adaptation Plan*, September 1, 2021b.

U.S. Department of Defense, *Department of Defense Climate Risk Analysis*, October 2021c.

U.S. Department of Defense, *DOD Dictionary of Military and Associated Terms*, November 2021d.

U.S. Department of Defense, *Mission Assurance Construct*, DoD Instruction 3020.45, May 2, 2022a.

U.S. Department of Defense, *2022 National Defense Strategy of the United States of America*, October 27, 2022b.

U.S. Department of Transportation, Federal Highway Administration, "Vulnerability Assessment and Adaptation Framework, 3rd Edition," webpage, last updated January 22, 2018. As of June 3, 2022: https://www.fhwa.dot.gov/environment/sustainability/resilience/adaptation_framework/chap00.cfm

U.S. Geological Survey, "What Is the Difference Between Weather and Climate Change?" webpage, undated. As of July 21, 2022:
https://www.usgs.gov/faqs/what-difference-between-weather-and-climate-change

U.S. Government Accountability Office, *Military Training: Compliance with Environmental Laws Affects Some Training Activities, but DoD Has Not Made a Sound Business Case for Additional Environmental Exemptions,* GAO-08-407, March 2008.

U.S. Government Accountability Office, *Defense Infrastructure: DOD Should Better Manage Risks Posed by Deferred Facility Maintenance,* GAO-22-104481, January 2022.

U.S. House of Representatives, Implications of Closing the Vieques Training Facility: Hearing Before the Committee on Armed Services, June 27, 2001.

Vergun, David, "Action Team Leads DOD Efforts to Adapt to Climate Change Effects," DoD News, April 22, 2021.